Manual Mobilization of the Joints

Volume I

The Extremities

Manual Mobilization of the Joints

The Kaltenborn Method of Joint Examination and Treatment

Volume I

The Extremities

by Freddy M. Kaltenborn
in collaboration with
Olaf Evjenth,
Traudi Baldauf Kaltenborn,
Dennis Morgan, and Eileen Vollowitz

6th Edition 2002
Published and distributed by
Olaf Norlis Bokhandel
Oslo, Norway

Also distributed by
OPTP
Minneapolis, Minnesota, USA

Published 2002 and distributed by:
Olaf Norlis Bokhandel, Universitetsgaten 24, N-0162 Oslo, Norway

English edition also distributed by OPTP
PO Box 47009, Minneapolis, MN 55447, USA
(612) 553-0452; (800) 367-7393; Fax: (612) 553-9355

First Edition	*Manual Therapy for the Extremity Joints*, 1974
Second Edition	*Manual Therapy for the Extremity Joints*, 1976 Revised and translated by Barbara Robinson P.T.
Third Edition	*Mobilization of the Extremity Joints*, 1980 Revised and translated by Dennis Morgan, P.T.
Fourth Edition	*Manual Mobilization of the Extremity Joints*, 1989 Edited and translated by Dennis Morgan, D.C., P.T.
Fifth Edition	*Manual Mobilization of the Joints, Volume I: The Extremities*, 1999 Revised and edited by Dennis Morgan, D.C., P.T. and Eileen Vollowitz, P.T.
Sixth Edition	*Manual Mobilization of the Joints, Volume I: The Extremities*, 2002 Revised and edited by Eileen Vollowitz, P.T.

This book is a companion volume to
Manual Mobilization of the Joints, Volume II, The Spine, 2002 (ISBN 82-7054-052-8)

Also published in the following languages:

Volume I: The Extremities

Finnish	1986
German	1972, 2002
Greek	2001
Japanese	1988
Korean	2001
Norwegian	1960, 1993 (out of print)
Polish	1996
Portugese	2001
Spanish	1986, 2001

Volume II: The Spine

Chinese	2000
German	1972, 2002
Greek	2001
Japanese	1988
Korean	2001
Norwegian	1964, 1989
Polish	1998
Spanish	2000

Printed in the United States of America by Banta ISG, Minneapolis, MN

ISBN 82-7054-043-9

Foreword to third edition

Professor Kaltenborn has honoured me by asking me to write a Foreword to this book. I do so gladly for two reasons: our long friendship, and the nature and scope of his work.

For some 60 years I have watched the expansion of orthopaedic surgery after the First World War, and that of physical (or orthopaedic) medicine after the Second. Manual Therapy is the child of both, and is rapidly becoming a part of nonspecialized medicine and, of course, of physiotherapy. It is of special value to the rural practitioner, who can now often treat his or her patients in a way that would formerly have needed a journey to some distant town at considerable expense in time and money. Its value for physiotherapists needs no stressing.

The book describes each test and mobilization in simple but precise language, language reinforced by the numerous clear photographs. What I especially like about it is its marriage of functional anatomy to clinical practice. I do not say this because of the references to part of my own work in arthrology; and ask the reader to believe me. Yet, after all, the most important part of anatomical research and teaching is that which can be used for the relief of human ills.

To comfort always, to alleviate often, to cure sometimes: these are the three aims of the healer. Professor Kaltenborn will help them to be achieved in fuller measure.

I salute his book and say to it:

Go, little book; and, in each distant part
Whereto you go, enlarge the healer's art!

M.A. MacConaill, 1980
Professor in Anatomy, Cork, Ireland
Referenced in *Gray's Anatomy: Arthrology*, 1980

Acknowledgments

Each time I publish a new edition of this book I say it will be my last. With each publication I believe I have finally and fully explained the basis for my clinical thinking. Yet, the more I write, the more my colleagues and students find in my writings to discuss and question. Concepts I assumed were generally agreed upon and understood, I discover instead to be sources of discord and confusion. And so, I must write again, and again, each time expanding and clarifying what seems so simple in patient treatment, yet has been so difficult to explain in words. I am grateful for this continued discourse, as it has made me both a better teacher and a better practitioner.

This book would not exist were it not for the interest and efforts of my collaborators and colleagues. Many spent long hours in discussions and careful editing of my drafts. For this I would especially like to thank Olaf Evjenth, Bjørn Støre, and Jochen Schomacher. I am indebted to Eileen Vollowitz, P.T., for her patience, insight and talent to change my brief notes into cohesive prose — again, and again. And I thank my many colleagues and former students from all over the world who made many valuable suggestions.

Last but not least, I thank my wife Traudi Baldauf Kaltenborn for her love and support over the last 30 years. She has been invaluable in providing professional advice and technical support. She brought me calm and focus when I most needed it. I could never have accomplished so much without her.

Freddy M. Kaltenborn, 2002

About the author

Kaltenborn's career began as a physical educator and athletic trainer in Germany in 1945 and as a physical therapist in Norway in 1949. He apprenticed with Dr. James Mennell and Dr. James Cyriax in London, England, from 1952 to 1954 to learn more about orthopedic medicine, and received his certification to teach the Cyriax approach in 1955. Thereafter he studied at the British School of Osteopathy. Upon return to his native Norway, Kaltenborn worked to incorporate these concepts into his own system.

In 1958 Kaltenborn was certified in chiropractic by the Forschungs- und Arbeitsgemeinschaft für Chiropraktik (FAC) in Germany and taught chiropractic to the medical doctors within FAC between 1958 and 1962. By 1962 the FAC had incorporated the Kaltenborn Method into their approach and changed the name of their professional practice from "Chiropraktik" to "Chirotherapy." Kaltenborn continued to instruct FAC practitioners until 1982.

In 1962 Kaltenborn studied at the London College of Osteopathy in London, England and subsequently was approved as an osteopathic instructor by Dr. Alan Stoddard in 1971. Kaltenborn was certified in orthopedic manipulative therapy by the International Seminar of Orthopaedic Manipulative Therapy (ISOMT) in 1973. Between 1977 and 1984 he served as a professor at the Michigan State University, College of Osteopathic Medicine, USA.

Kaltenborn practiced physical therapy in his native Norway for thirty-two years, from 1950 to 1982. During that time he instructed countless physical therapists, medical doctors, and many osteopaths and chiropractors in manual treatment methods. He introduced manual therapy to Norwegian physical therapists and was instrumental in developing manual therapy education and certification standards there. Together with Norwegian medical doctors, Kaltenborn also brought the benefits of manual therapy to the attention of the Norwegian national health care system, which by 1957 had recognized the effectiveness of manual therapy by reimbursing skilled manual therapy services at twice the rate of other physical therapy treatments.

Throughout his professional career, Professor Kaltenborn campaigned tirelessly for the creation of international educational standards and certification in manual therapy. He was a founding member of the International Federation of Orthopaedic Manipulative Therapists (IFOMT), now a subgroup of the World Confederation of Physical Therapists (WCPT). Professor Kaltenborn contributed to the creation of IFOMT's first manual therapy education and certification standards, the first such standards to be recognized by an international professional organization.

About this book

I wrote this book in the belief that joint mobilization skills are useful for almost all physical therapists and should be taught in all physical therapy schools.

Both this textbook and its companion volume, *Manual Mobilization of the Joints: The Spine (2002)*, are intended for beginning students. These books are also recommended for professionals who did not receive Kaltenborn-Evjenth System Orthopedic Manual Therapy (OMT) basic training as part of their professional education and must now learn these basic skills through continuing education courses and residencies.

This book concentrates on basic manual, passive extremity joint evaluation and mobilization techniques with an emphasis on the application of biomechanical principles. Keep in mind that these joint mobilization techniques are but one part of OMT Kaltenborn-Evjenth System practice (see *OMT Overview*, page 10). There are many texts available covering other areas of OMT practice, including soft tissue mobilization techniques, stabilization techniques for joint hypermobility, more advanced joint mobilization techniques, and much more.

New in this edition

Advanced mobilization progressions outside the resting position, including pre-positioning up to the point of restriction, are presented with clear guidelines for progression of technique.

Grades of translatoric movement, interpreted and applied based on manual movement quality assessment, are graphically illustrated and detailed. I also note the most effective grade of movement for the application of each technique. The ability to palpate and interpret grades of movement, a concept I first introduced in 1952, separates the master practitioner from the novice.

Guidelines to reduce pain and inflammation and to relax muscles using gentle Grade I and II mid-range joint mobilizations are more thoroughly explained than in earlier editions.

Techniques for longer duration joint mobilization procedures utilize alternate grips with improved leverage, fixation, mobilization straps, and therapist body mechanics. Grade III stretch-mobilizations are most effective when applied for longer durations. In this edition we further clarify how best to apply these techniques.

Table of contents

PRINCIPLES

TECHNIQUE

APPENDIX

OMT Kaltenborn-Evjenth System

Orthopedic medicine specializes in the diagnosis and treatment of musculoskeletal conditions.[1] The physical therapy specialty Orthopedic Manipulative Therapy (OMT) is an important part of orthopedic medicine. Much of OMT is devoted to the evaluation and treatment of joint and related soft tissue disorders and one of the primary treatment methods is mobilization. When examination reveals joint dysfunction, especially decreased range of motion (i.e., hypomobility), the joint mobilization techniques described in this book are often effective.

The OMT Kaltenborn-Evjenth system is the result of many years of collaboration between physical therapists and physicians, first in the Nordic countries from 1954 to 1970, and then worldwide. The system began in 1954 with joint testing and treatment only and was known as "Manual Therapy ad modum Kaltenborn." It later became known as the Norwegian System or the Nordic System. In 1973, Olaf Evjenth and I began our decades long collaboration to develop the system as we know it today, the OMT Kaltenborn-Evjenth System.

The Orthopedic Manual Therapy (OMT) Kaltenborn-Evjenth System is a physical therapy treatment approach based on information and experience from sports medicine, traditional physical therapy, osteopathy, orthopedic medicine, and the further innovations of the many therapists who have practiced manual therapy techniques. The methods presented in this book focus primarily on manual joint testing and treatment, and are but one part of the OMT Kaltenborn-Evjenth System.

Development of the OMT Kaltenborn-Evjenth System

The story of the OMT Kaltenborn-Evjenth approach to manual therapy began in the 1940s when I became frustrated in my attempts to treat patients with spinal disorders. First as a physical educator treating disabled soldiers in 1945 and later as a physical therapist in 1949, I found that the massage combined with mobilization and manipulation (especially for the extremities) I had learned from physical education, along with the active and passive

1 In recent years, orthopedic medicine has become known as "manual medicine" or "musculoskeletal medicine."

movements I had learned from conventional physical therapy training, was limited in its effectiveness. Many of the spinal patients I was unable to help reported finding relief from chiropractic treatment.

In Norway at that time doctors of physical medicine would only support the introduction of a new physical therapy approach if it came from within the traditional practice of medicine. Therefore, I turned to the work of Dr. James Mennell, a physician of physical medicine, and Dr. James Cyriax, a physician of orthopedic medicine, both at St. Thomas Hospital in London. These physicians were unusual in their commitment to bringing their experience in manual medicine to the training of physical therapists. Mennell began teaching his techniques to physical therapists as early as 1906 and wrote his first textbook for physical therapists in 1917, *Physical Treatment by Movement and Massage* (published by Churchill, London). He later published *The Science and Art of Joint Manipulation, Volume I: The Extremities* (1949) and *Volume II: The Spine* (1952). Dr. Cyriax's 1947 *Textbook of Orthopaedic Medicine, Volume I: Diagnosis* and *Volume II: Treatment* remain basic texts on evaluating and treating soft tissue disorders for OMT Kaltenborn-Evjenth system training today.

In the early 1950's I went to London with my colleague R. Stensnes, to observe the joint mobilization techniques of Dr. Mennell and to study with Dr. Cyriax. Upon my return to Norway, I demonstrated my newly acquired skill at the Medical Association for Physical Medicine, which then agreed to sponsor my first course on Cyriax's approach. The course was taught to eight physical therapists in 1954 and was cosponsored by the Physical Therapy Association of Oslo. This signalled the beginnings of a significant change in the Norwegian medical establishment's view of manual therapy. Well into the 1950's, many Norwegian physicians still considered manual therapy outside of the practice of medicine and therefore did not support its practice by physical therapists or by medical doctors. Norwegian physician Eiler Schiøtz documented manual therapy's historical place in medicine in his monograph, the *History of Manipulation* (1958), and so helped support the eventual inclusion of manual therapy within the scope of traditional medical practice in Norway.

In 1955, Dr. Cyriax visited Norway to approve courses in his approach and to instruct and examine the first physical therapists to complete those studies. These graduates formed the Norwegian Manipulation Group, an ongoing study group that practiced and further developed what was becoming a specialized OMT approach for physical therapists.

Up to this point, only regional, nonspecific approaches to evaluating and treating spinal patients were used by Mennell, Cyriax, and the Norwegian Manipulation Group. But Alan Stoddard, M.D. and D.O., was performing more specific techniques within the practice of osteopathy to treat the spine. Stoddard describes these techniques in his textbooks, *Manual of Osteopathic Technique* (1959) and *Manual of Osteopathic Practice* (1969), which made osteopathic techniques more accessible to physical therapists and medical doctors.

In the late 1950's and early 1960's, I studied at both schools where Stoddard was an instructor: The British School of Osteopathy and The London College of Osteopathy.

With Stoddard, I brought selected osteopathic techniques to the Norwegian Manipulation Group. Cyriax and Stoddard worked with me for many years to determine which evaluative and treatment tools from physical therapy, sports medicine, orthopedic medicine, and osteopathy would most benefit physical therapy practice and should be a part of manual therapy training for physical therapists.

■ MT ad modum Kaltenborn

1958 - 1972

I began to develop my own theories and techniques and to incorporate these into our evolving OMT system. My integrated approach became known as "Manual Therapy (MT) ad modum Kaltenborn" or "The Kaltenborn Method."[2]

Among my contributions were an emphasis on translatoric joint play movements in relation to a treatment plane for evaluating and mobilizing joints, the use of grades of movement, the convex-concave rule, three-dimensional pre-positioning for joint movement, protecting adjacent nontreated joints during procedures, self-treatment, and ergonomic principles applied to protect the therapist. (See *Special Features*, page 7.)

During this period of time, my method included:

» Orthopedic Medicine (from J. H. Cyriax and J. B. Mennell)

» Osteopathy (from A. T. Still and A. Stoddard)

» My original techniques (F. M. Kaltenborn)

I emphasized functional evaluation of the locomotor system and the biomechanical treatment of dysfunction. In those days patients often presented with joint stiffness due to prolonged immobilization

---

2 In 1958, only Norwegian P.T.'s referred to my approach as "Manual Therapy ad modum Kaltenborn." During the 1960's practitioners in other European countries adopted the term as well, as did Nordic medical doctors in 1965.

in plaster casts for the treatment of fractures and dislocations. (Modern-day treatment of these disorders incorporates joint movement to prevent such secondary joint problems.) My methods supplemented traditional physical therapy approaches with treatment techniques for:

» Symptom relief, especially for pain.
» Relaxation of muscle spasm.
» Stretching of shortened joint and muscle connective tissues.

From 1960, I presented my MT courses to physical therapists from the Nordic countries. From 1962 physicians attended as well. At this time, Dr. Schiøtz and other Scandinavian physicians created the Nordic Physicians Manual Medicine Association (NFMM). The association also developed groups to teach my MT system and named educational coordinators for Denmark, Norway, Finland and Sweden, for which I served as Nordic Educational Director for Physicians and Physical Therapists. As practicing physicians, the NFMM members reported their clinical results on integrating this manual treatment approach into their practices, and thus contributed to the fine-tuning of the system.

■ OMT Kaltenborn-Evjenth System

1973 - present

Olaf Evjenth, a skilled Norwegian practitioner with a background in physical education, athletic training and physical therapy, joined me in 1958. He expanded my approach with specialized techniques for muscle stretching and coordination training. In particular, he believed in more intensive training for patients and developed programs that, in addition to monitoring pain and range of movement, assessed performance. Evjenth also modified specific exercises for patient use at home with automobilization, autostabilization, and autostretching. Evjenth and I, together with members of the Norwegian Manual Therapy Group, began to develop and use additional self-treatment techniques, equipment for home treatment, and ergonomic innovations including mobilization wedges, fixation belts, and grips to make treatments more effective and less physically stressful for the therapist (always a concern in our system).

In 1990, Evjenth introduced symptom alleviation testing as a method for localizing lesions and improved symptom provocation testing. This aided in making evaluations more specific. He also improved techniques for protecting adjacent nontreated joints during manual mobilization procedures.

Figure I-3: Evjenth (left) and Kaltenborn in Canada in 1968, introducing our OMT system to North America

Multiple treatment techniques, often performed within the same treatment session, are basic to our system. This approach to treatment was improved further as Evjenth and I began to sequence techniques for the most effective results.

We presented the "OMT Kaltenborn-Evjenth Concept" worldwide in 1973, when Evjenth and I joined Cyriax, Hinsen, and Stoddard to found the International Seminar of Orthopaedic Manipulative Therapy. At that time we included:

» MT ad modum Kaltenborn
» Contributions from Olaf Evjenth
» Contributions from other practitioners

My philosophy has always been to integrate useful tools from other approaches. Over the years, the Kaltenborn-Evjenth OMT system benefited from the contributions of many physical therapists and physicians, both in the Nordic countries and worldwide. A few have been especially important to our approach and should be mentioned here: Herman Kabat, M.D. and physical therapists Margaret Knott and Dorothy Voss developed the proprioceptive neuromuscular facilitation (PNF) principles behind our active relaxation and muscle reeducation techniques; Oddvar Holten P.T. developed medical training therapy (MTT) and Dennis Morgan D.C., P.T., developed specialized exercise training programs and equipment which we now incorporate into our OMT treatment programs; Geoffrey Maitland of Australia, with whom I have had many stimulating discussions about our concepts and approach. Many other practitioners also had an influence on my thinking, including S.V. Paris, R. McKenzie, M. Rocabado, B. Mulligan and others.

In 1974, Maitland (of Australia) and I, together with therapists trained in both our OMT system and Maitland systems, founded the International Federation of Orthopaedic Manipulative Therapy (IFOMT), which later became a subgroup of the World Confederation of Physical Therapists. Through IFOMT's international forums, OMT Kaltenborn-Evjenth system representatives have been a major influence on physical therapy. Our system's continuing evolution has been aided by this opportunity for its practitioners and founders to interact with representatives of other OMT approaches worldwide.

OMT Kaltenborn-Evjenth system standards formed the basis for IFOMT educational and certification standards adopted in 1974 and 1975, which must be met by all participating members. Many other countries in which the OMT Kaltenborn-Evjenth system is taught are beginning to develop similar educational and certification standards. To date, our system is taught in the Nordic countries, in Australia, Austria, Belgium, France, Germany, Greece, Italy, Japan, Korea, Netherlands, Poland, Spain, Switzerland, and in North and South America.

Today, our system has expanded to encompass evaluation, treatment and research for a complete neuro-musculoskeletal approach to manual physical therapy. Education incorporates clinically supervised residencies and written and practical examinations. At the highest levels of training, practitioners are also required to conduct independent research in the field of manual therapy.

Special features

As the OMT Kaltenborn-Evjenth System more extensively influences the practice of physical therapy, so our system continues to evolve. But certain special features can be identified as basic and unique in their application to our system. In many cases we were the first to introduce these concepts to physical therapy practice, which are now widely accepted and practiced.

Biomechanical approach to treatment and diagnosis

Manipulative technique has changed over the past 50 years. Traditional manipulations applied long-lever rotational movements. The compressive forces produced by these long-lever rotational movements sometimes injured joints.

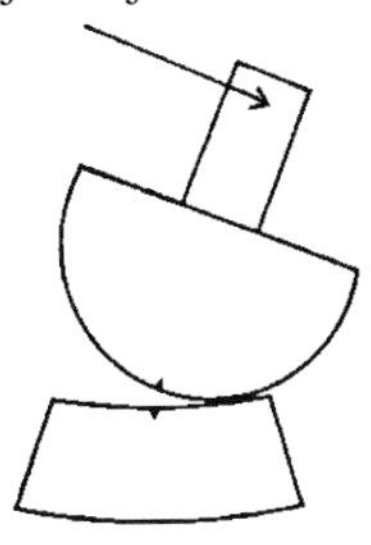

Figure I.4
Prior to 1952, practitioners used long-lever rotation techniques (passive continuation of active movement)

In the 1940s, James Mennell, M.D. introduced shorter lever rotational manipulations which reduced the possibility of joint damage. In 1952 Norwegian manual therapists adopted these short-lever manipulative techniques.

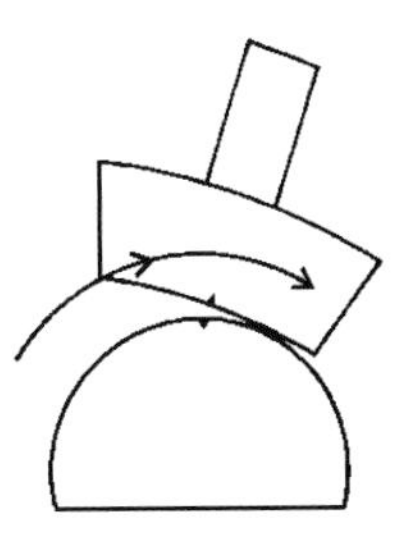

Figure I.5
In 1952, practitioners began to use short-lever rotation techniques

In 1954, I introduced the concept of translatoric bone movements, in the form of linear translatoric traction and gliding *in relation to a treatment plane*, to further reduce joint compression forces. Over the next 30 years I worked to incorporate translatoric joint movements into a comprehensive joint evaluation and treatment approach that reduced the need for short-lever rotation mobilizations. By 1979, Evjenth and I had refined our techniques to eliminate rotatory forces in extremity joint treatment, and by 1991, had accomplished the same for spinal manipulations.

In the OMT Kaltenborn-Evjenth System, biomechanical principles form the core of the analysis and treatment of musculoskeletal conditions.

» Translatoric treatment in relation to the Kaltenborn Treatment Plane allows for safe and effective joint mobilization.

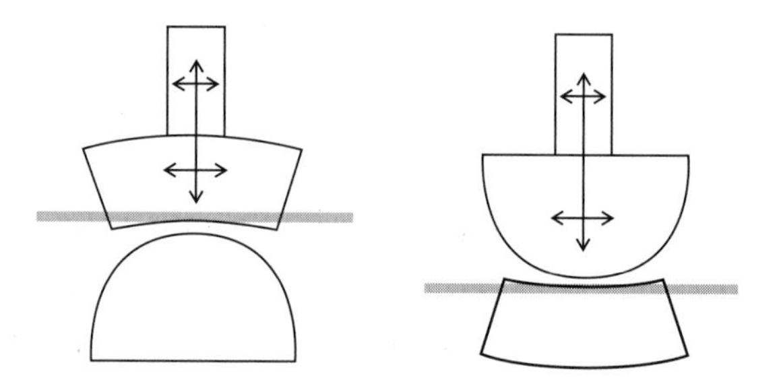

Figure I.6
In 1954, I incorporated the concept of translatoric bone movement in relation to the treatment plane

» The therapist evaluates the translatoric joint play movements of traction and gliding by feeling the amount of slack in the movement and sensing the end-feel. The therapist uses grades of movement to rate the amount of joint play movement they palpate.

» Three-dimensional joint positioning, carefully applied before a mobilization, refines and directs the movement.

» The Kaltenborn Convex-Concave Rule allows indirect determination of the direction of decreased joint gliding to insure normal joint mechanics during treatment.

» The therapist evaluates and treats all combinations of movements, coupled and non-coupled.

» The therapist uses specific evaluation and specific treatment, including special tests to localize symptomatic structures, and to treat hypermobility in addition to hypomobility.

Combination of techniques

The use of multiple treatment techniques, often in one treatment session, has always been part of our system. For example, techniques to improve joint mobility are often preceded by pain-relief and soft-tissue-mobilization techniques such as functional massage and muscle stretching. Self-treatment is an important part of our system and may include instruction in automobilization, autostretching, autotraction, strengthening, stabilization, or coordination exercises. Advice on body mechanics and ergonomics is important to maintain improvements gained in therapy and to prevent recurrences.

Trial treatment

An experienced practitioner views any treatment procedure also as an evaluation procedure. I formalized this concept within my system in 1952, with the term "trial treatment," where the manual therapist confirms the initial physical diagnosis with a low-risk trial treatment as an additional evaluation procedure.

Ergonomic principles for the therapist

The OMT Kaltenborn-Evjenth System emphasizes good *therapist* body mechanics. An example of this was my development in the 1950's of the first pneumatic high-low adjustable treatment table designed for manual physical therapy practice. Our practitioners have since developed a number of treatment techniques and tools for efficiency and safety, including mobilization and fixation belts, wedges, and articulating tables.

Overview

OMT Kaltenborn-Evjenth System for Physical Therapists

The Kaltenborn Method for joint testing and mobilization presented in this book is part of the larger scope of OMT Kaltenborn-Evjenth System practice.

I. Physical Diagnosis (biomechanical and functional assessment)

A. **Screening exam:** An abbreviated exam to quickly identify the region where a problem is located and focus the detailed examination

B. **Detailed exam:**

1. ***History***: Narrow diagnostic possibilities; develop early hypotheses to be confirmed by further exam; determine whether or not symptoms are musculoskeletal and treatable with OMT. *(Includes present episode, past medical history, related personal history, family history, review of systems)*

2. ***Inspection***: Further focus the exam. *(Includes posture, shape, skin, assistive devices)*

3. ***Tests of function***

   a. **Active and passive movements:** Identify location, type, and severity of dysfunction. *(Includes standard-anatomical-uniaxial movements and combined-functional-multiaxial movements)*

   b. **Translatoric joint play movements:** Further differentiate articular from nonarticular lesions; identify directions of joint restrictions. *(Includes traction, compression, gliding)*

   c. **Resisted movements:** Test neuromuscular integrity and status of associated joints, nerves and vascular supply.

   d. **Passive soft tissue movements:** Differentiate joint from soft tissue dysfunction and the type of soft tissue involvement. (*Includes physiological movements, accessory movements)*

   e. **Additional tests** *(Includes coordination, speed, endurance, functional capacity assessment ...)*

4. ***Palpation*** *(Includes tissue characteristics, structures)*

5. ***Neurologic and vascular examination***

C. **Medical diagnostic studies** *(Includes diagnostic imaging, lab tests, electro-diagnostic tests, punctures)*

D. **Diagnosis and trial treatment**

II. Treatment

A. To relieve symptoms (most often pain)

1. ***Immobilization***
   - *General: bed rest*
   - *Specific: corsets, splinting, casting, taping*
2. ***Thermo-Hydro-Electro (T-H-E) therapy***
3. ***Pain relief joint mobilization*** *(Grade I-II Slack Zone in the actual resting position)*
   - *Intermittent manual traction*
   - *Vibrations, oscillations*
4. ***Special procedures*** *(Includes acupuncture, acupressure, soft tissue mobilization ...)*

B. To increase mobility

1. ***Soft tissue mobilization***
   - a. **Passive soft tissue mobilization**
     - *Classical, functional, and friction massage*
   - b. **Active soft tissue mobilization**
     - *Contract-relax, reciprocal inhibition, muscle stretching*
2. ***Joint mobilization***
   - a. **Relaxation joint mobilization** (Grade I - II)
     - *Three-dimensional, prepositioned mobilizations*
   - b. **Stretch joint mobilization** (Grade III)
     - *Manual mobilization in the joint (actual) resting position*
     - *Manual mobilization at the point of restriction*
   - c. **Manipulation**
     - *High velocity, short amplitude, linear thrust movement*
3. ***Neural tissue mobilization***
   To increase mobility of dura mater, nerve roots, and peripheral nerves
4. ***Specialized exercise***
   To increase or maintain soft tissue length and mobility and joint mobility

C. To limit movement

1. ***Supportive devices***
2. ***Specialized exercise***
3. ***Treatments to increase movement in adjacent joints***

D. To inform, instruct, and train

Exercises and education to improve function, compensate for injuries, and prevent reinjury. Instruction in relevant ergonomics and self-care techniques, e.g., medical training therapy, automobilization, autostabilization, autostretching, back school, etc.

III. Research

Clinical trials to determine the efficacy of various single and combined treatment methods

Notes

PRINCIPLES

1 Extremity joint movement

Joint anatomy

Articular surfaces

Classical descriptions of joint surfaces as "plane" or "spheroid" are terms of convenience and not completely accurate. No joint surface is perfectly flat or part of a cylinder, cone, or sphere. In reality, all joint surfaces have a certain amount of curvature, which is not constant but changes from point to point. MacConaill's classification of joint surfaces more accurately reflects this reality. He describes joint surfaces as either ovoid or sellar.

Ovoid joint surfaces (Figure 1.1a) can be either convex or concave in all directions and are similar to a piece of egg shell, in that their surfaces are of a constantly changing angular value.

Sellar or saddle surfaces (Figure 1.1b) are inversely curved with convex and concave surfaces situated at right angles to each other.

Figure 1.1
Classification of joint surfaces (after MacConaill)

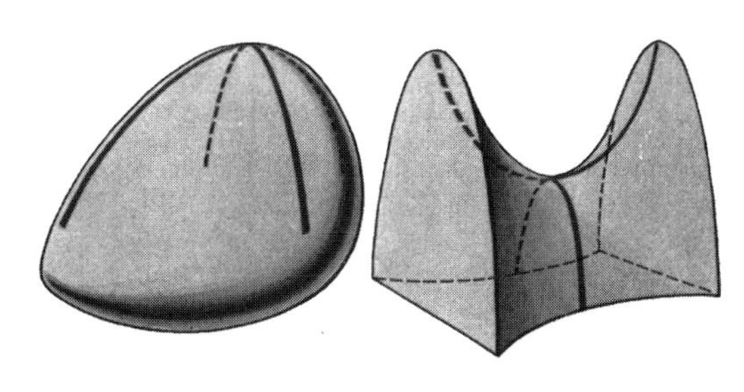

a. Ovoid surface *b. Sellar surface*

Bony connections

In most joint positions the articular surfaces are not fully congruent. The incongruence of joint partners is due to the differences in curvature of the articular surfaces, e.g., the convex partner is more curved (smaller radius of curvature) than its concave joint partner.

Joints have traditionally been classified only by their morphology and mechanical characteristics. In more recent years, MacConaill developed a more useful classification of joints based on the type of bone movement allowed at each joint. In the following pages we review both.

A conventional classification of joints

Bony connections are conventionally classified according to their morphology. Synovial joints are further classified according to their mechanical characteristics.

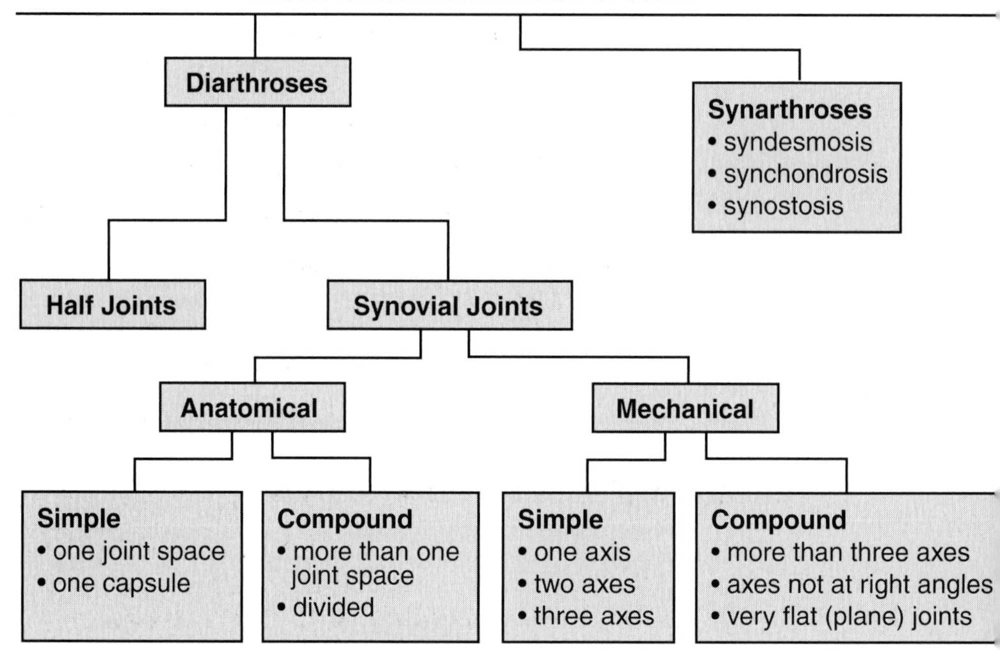

Bony connections are first classified as either synarthroses or diarthroses.[1]

Synarthroses are named according to the type of tissue that connects them:

» Syndesmosis: fibrous tissue

» Synchondrosis: cartilage

» Synostosis: bone

Diarthroses are classified as synovial and half joints (e.g., symphysis, uncovertebral joints).

» Synovial joints with less than 10 degrees of movement are called amphiarthroses.

1 Since most diarthroses are synovial joints and there are very few half joints in the human body, many people use the term "synovial joints" synonymously with "diarthroses."

Synovial joints are classified as anatomical or mechanical, simple or compound joints.

» **Anatomically simple** joints have only one joint space.

» **Anatomically compound** joints have:

- More than one joint space, divided by a meniscus or articular disc.

» **Mechanically simple** joints have one, two or three axes which are situated at right angles to each other:

- One Axis: ginglymus (hinge) and trochoid (pivot)
- Two axes: ellipsoid, sellar (saddle)
- Three axes: spheroid (enarthroses, ball and socket)

» **Mechanically compound** joints have:

- More than three axes
- Axes not situated at right angles to each other
- Very flat joint surfaces without the usual axes

MacConaill's classification of joints

MacConaill describes four structural classifications of synovial joints which are correlated with the types of bone movements and the degrees of freedom allowed at each articular pair:

» **Unmodified ovoid**: (art. spheroidea), ball and socket, triaxial, e.g., hip and shoulder joints

» **Modified ovoid**: (art. ellipsoidea), ellipsoid, biaxial, e.g., metacarpophalangeal (MCP) joints

» **Unmodified sellar**: (art. sellaris), saddle, biaxial, e.g., first carpometacarpal joint

» **Modified sellar**: (art. ginglymus), hinge, uniaxial, e.g., interphalangeal joints

Reference positions

Anatomical planes of reference

The body is traditionally divided into three anatomical (cardinal) planes that are situated at right angles to each other and intersect at the body's center of gravity. These planes of reference are used for describing and measuring anatomical bone movements.

The median plane divides the body symmetrically into right and left halves and all planes parallel to this are called *sagittal* planes. Planes dividing the extremities into right and left halves are called *dorsal-ventral, dorsal-palmar,* or *dorsal-plantar* planes.

The frontal plane divides the body into anterior (ventral) and posterior (dorsal) halves. Planes dividing the extremities into anterior and posterior halves are called *medial-lateral, radial-ulnar,* or *tibial-fibular* planes.

The transverse plane or horizontal plane divides the body into *cranial* and *caudal* halves and the extremities into *distal* and *proximal* halves.

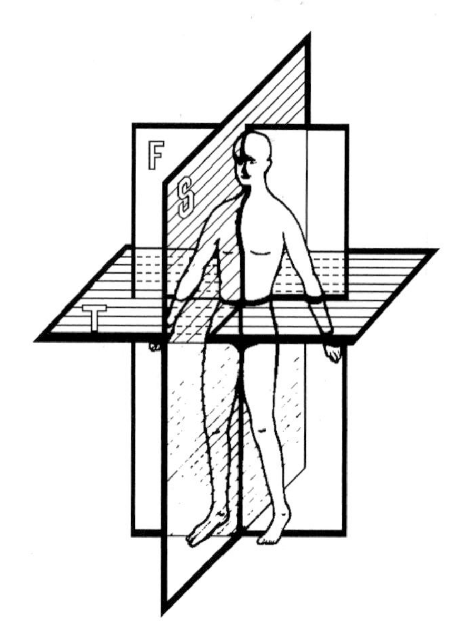

Figure 1.2 Anatomical planes of reference

Anatomical axes of reference

The anatomical axes lie at the intersection of two anatomical planes and anatomical bone movements take place around these axes.

The **frontal axis** lies at the intersection of the frontal and transverse planes and runs from right to left. In the extremities, this axis is called *transverse, medial-lateral, radial-ulnar*, or *tibial-fibular* (Figure 1.3a).

The **sagittal axis** lies at the intersection of the sagittal and transverse planes and runs in a dorsal-ventral direction. In the extremities, this axis is called *dorsal-ventral, dorsal-palmar, dorsal-plantar*, or *posterior-anterior* (Figure 1.3b).

The **longitudinal (vertical) axis** lies at the intersection of the sagittal and frontal planes and runs in a cranial-caudal direction. In the extremities, this axis passes through a part of a bone such as the neck of the femur or the entire length of a bone e.g., the shaft of the humerus, clavicle, etc. (Figure 1.3c, longitudinal axis of the humerus).

Figure 1.3
Anatomical axes
(after MacConaill)

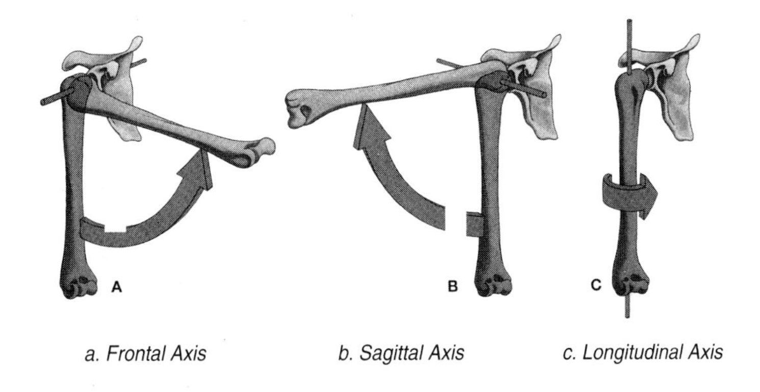

a. Frontal Axis b. Sagittal Axis c. Longitudinal Axis

During active and passive movements the mechanical axis does not remain stationary, due to the joint's changing radius of curvature. The constantly changing curvature and the lack of congruency allows roll-gliding to occur in all joints. Since the axis for movement does not remain stationary, we use the term *instantaneous axis of rotation* (IAR) to emphasize this fact. Normally the IAR is located on the convex side of the joint.

Three-dimensional joint positioning

The effectiveness of joint evaluation and mobilization treatment can be enhanced by placing the joint specifically in one, two, or three planes. For practical purposes, we classify joint positions into five categories:

» Zero position

» Resting position (Loose-packed position)

» Actual resting position

» Nonresting positions

» Close-packed position

Zero position

All joint range of motion measurements are taken from the zero starting position, if possible.[1] The range of motion is measured with a goniometer on both sides of zero. For example, a movement of thirty degrees flexion and ten degrees extension is written: flexion/extension 30-0-10. However, if there is limitation of movement, with movement only possible on the flexion side of zero, both figures are written on the left side of zero as in flexion/extension 30-10-0.

Resting position

The resting position (loose-packed position) is the position (usually three-dimensional) where periarticular structures are most lax, allowing for the greatest range of joint play.[2] With many joint conditions, this position is also the patient's position of comfort (symptom-relieving posture) affording the most relaxation and least muscle tension.

1 This book uses the internationally accepted zero position originally described by Cave and Roberts in 1936, and later by Chapchal (1957) and Debrunner (1966).

2 MacConnail referred to the "resting position" as the "loose packed position."

The resting position is useful for:

» evaluating joint play through its range of motion, including end-feel.

» treating symptoms with Grade I-II traction-mobilization within the slack.

» treating hypomobility with Grade II relaxation-mobilization or Grade III stretch-mobilization.

» to minimize secondary joint damage due to long periods of immobilization associated with casting and splinting.

To find the resting position:

1) Position the joint in the approximate resting position according to established norms. For example, resting position for the hip is approximately at 30° flexion/30° abduction/slight external rotation.

2) In this approximate resting position, apply several gentle Grade II traction joint play tests to the first stop, feeling for the ease and degree of movement.

3) Reposition into slightly more or less *flexion or extension* and apply the traction tests again until you locate the position with the greatest ease and degree of movement. Maintain this position as you proceed to the next step.

4) Repeat the traction tests with subtle repositionings into more or less *abduction* and apply the traction tests again until you locate the position with greatest ease and degree of movement. Maintain this flexion/extension and abduction/adduction position as you proceed to the next step.

5) Repeat the traction tests with subtle repositionings into more or less *rotation* until you find the position with the greatest ease and range of movement in <u>all three dimensions</u>. This is the resting position.

The resting position may vary considerably among individuals.

■ Actual resting position

The actual resting position is used in special circumstances where it is impossible, difficult, or impractical to use the true resting position, for example in the presence of intra- or extra-articular pathology or pain. In this case, the joint is positioned where the therapist notes the least soft tissue tension and where the patient reports least discomfort. This *momentary* or *actual resting position* is then used for initial evaluation and treatment.

You will determine the actual resting position using the same techniques for finding the resting position, looking for the joint position of greatest ease, greatest range of traction joint play, least muscle reactivity and least tissue tension in the area of the dysfunction. The actual resting position must also be where the patient reports least discomfort. Keep in mind that the actual resting position will display somewhat less ease and range than the resting position.

■ Nonresting positions

Many subtle joint dysfunctions only become apparent when the joint is examined outside the resting position (nonresting position) and can only be treated in such positions. Other nonresting positions are used to specifically position soft tissues for movement or stretch.

Since nonresting positions allow less joint play, more skill is required to perform techniques safely in these positions. Novice practitioners applying stretch mobilizations in nonresting positions are more likely to overstretch tissues and cause injury. Stretch mobilization treatment in positions other than the resting position are considered "advanced" in our system and should be introduced to practitioners only after they demonstrate competence with resting position mobilizations.

■ Close-packed position

The close-packed position is characterized by the following criteria:

» The joint capsule and ligaments are tight or maximally tensed.

» There is maximal contact between the concave and convex articular surfaces (see Figure 1.4a).

» Articular surface gliding is maximally reduced and only slight separation with traction forces is possible.

Joint play testing and mobilization is difficult to perform at or near the close-packed position.

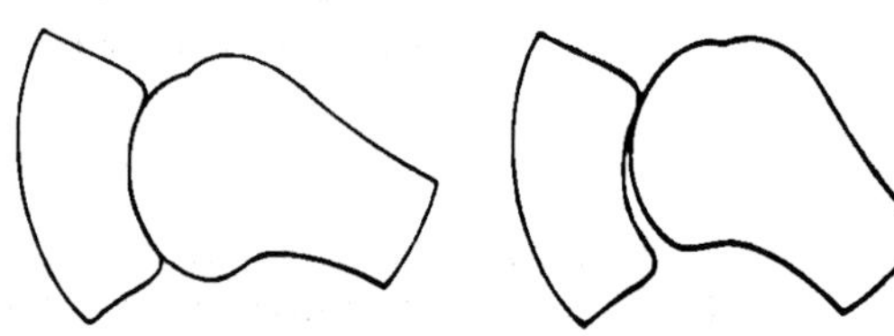

Figure 1.4a
Close-packed position

Figure 1.4b
Resting position

Bone and joint movement

Bone movements produce associated joint movements. The relationship between a bone movement (osteokinematics) and its associated joint movements (arthrokinematics) forms the basis for many orthopedic manual therapy (OMT) evaluation and treatment techniques.

Two types of bone movements are important in our OMT system:

Rotations: ***curved (angular) movement around an axis***

Translations: ***linear (straight-lined) movement parallel to an axis in one plane***[3]

Rotations of bone produce the joint movement of roll-gliding. Translations of bone result in the linear joint play movements of traction, compression, and gliding.

Bone movements	*Corresponding joint movements*
Rotatoric (angular) movement	**Roll-gliding**
- Standard (anatomical, uniaxial)	
- Combined (functional, multiaxial)	
Translatoric (linear) movement	**Translatoric joint play**
- Longitudinal bone separation	- Traction
- Longitudinal bone approximation	- Compression
- Transverse (parallel) bone movement	- Gliding

Rotations of a bone

Active movements occur around an axis and therefore, from a mechanical viewpoint, are considered rotations. All bone rotations can be produced passively as well. There are two types of bone rotations:

1) Standard, uniaxial bone movements (MacConaill's "pure, cardinal swing")

2) Combined, multiaxial bone movements (MacConaill's “impure arcuate swing")

3 From a mechanical perspective, translations can be curved or linear. Only linear translations are relevant to OMT practice. In this text, the term "translation" refers to *linear* translations.

Standard bone movements

Standard bone movements are bone rotations occurring around one axis (uniaxial) and in one plane. Standard movement is called "anatomical" movement when the movement axis and the movement plane are in anatomical (or cardinal) planes.

Anatomical bone movements beginning at the zero position are useful for describing and measuring test movements. They provide a standardized method for communicating examination findings that can be reproduced by other health care professionals. Anatomical movements of the bones in the three cardinal planes are described below.

Sagittal plane movements around a frontal axis

» **Flexion from zero**: movement occurs with the help of the flexor muscles and begins from the zero position.

» **Extension to zero**: movement occurs with the help of the extensor muscles from a flexed position back to the zero position.

» **Extension from zero**: extension movement continues past the zero position.

» **Flexion to zero**: movement occurs with the help of the flexor muscles from the above described extended position.

» **Palmar** and **dorsiflexion** in the hand, and **plantar** and **dorsiflexion** in the foot, describe movements around a transverse axis.

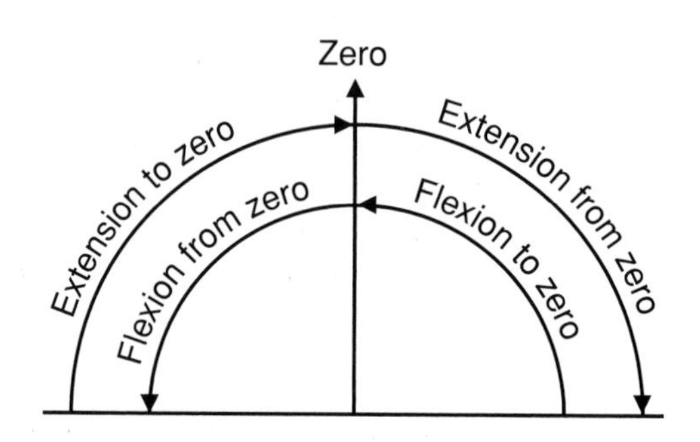

Figure 1.5
Sagittal plane movements

Frontal plane movements around a sagittal axis

» **Right and left side bending**: trunk or spinal movements occur in the frontal plane.

» **Abduction**: movements are away from the median or sagittal planes.

» **Adduction**: movements are towards the median or sagittal planes.

In the extremities, abduction and adduction movements are usually described relative to the regional anatomy, for example radial and ulnar flexion of the wrist.

Transverse plane movements around a longitudinal axis

The term **rotation** can be used to describe movement of a bone around its longitudinal (vertical) axis or an axis parallel to a longitudinal axis. A similar movement, **torsion** describes movement of bones in relationship to one another around an axis approximately parallel to their longitudinal axis, for example pronation and supination of the forearm.

» **Right and left rotation**: trunk or spinal movements in the transverse plane

» **Medial and lateral rotation**: movement of the extremities around longitudinal axes of bones

Combined bone movements

Bone movement that occurs simultaneously around more than one axis (multiaxial) and in more than one plane is called combined, or functional, movement. For example, simultaneous knee extension with rotation is a combined movement. These movements do not occur purely in cardinal planes and around stationary axes, but rather in oblique or diagonal directions. Combined movements represent most of the movements we carry out during daily activities. Manual therapists often examine combined movements in order to reproduce a patient's chief complaint and to analyze mechanisms of injury.

Combined movements are further classified as coupled and noncoupled movements according to the degree and nature of movement ease possible when flexion or extension and rotation are combined in various ways. *Coupled movements* have the greatest ease (greatest range and softest end-feel), for example, knee extension with external rotation. *Noncoupled movements* have less ease (less range and a harder end-feel), for example, knee extension with internal rotation. These movement distinctions are primarily applicable to spinal motion and are covered in more detail in the book *Manual Mobilization of the Joints, Volume II: The Spine.*[4]

Joint roll-gliding associated with bone rotations

Joint roll-gliding

In a healthy joint, functional movement (bone rotation) produces joint roll-gliding. Roll-gliding is a combination of rolling and gliding movement which takes place between two joint surfaces. Relatively more gliding is present when joint surfaces are more congruent (flat or curved), and more rolling occurs when joint surfaces are less congruent.[5]

Rolling occurs when new equidistant points on one joint surface come into contact with new equidistant points on another joint surface. Rolling is possible between two incongruent curved surfaces (i.e., surfaces of unequal radii of curvature). As illustrated below, a convex surface can roll on a concave surface (Figure 1.6a) or vice versa (Figure 1.6b).

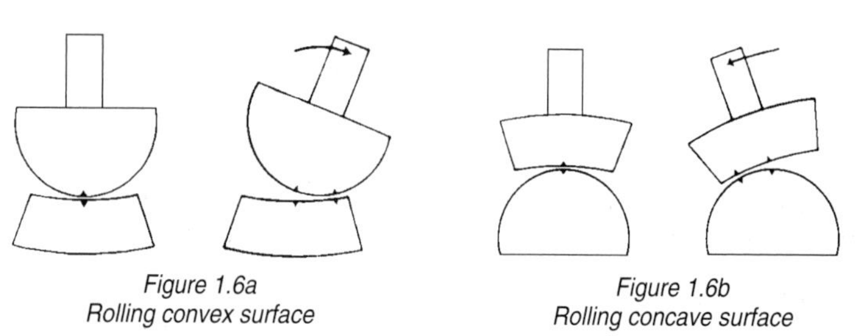

Figure 1.6a
Rolling convex surface

Figure 1.6b
Rolling concave surface

The **direction of the rolling component** of joint roll-gliding is always in the direction of the bone movement.

---

4 Terminology has changed as our concepts have evolved. Before 1992, coupled movement was called "physiological" movement and noncoupled movement was called "nonphysiological" movement. This older terminology was changed because "nonphysiological" movement was sometimes misinterpreted to mean abnormal movement, when in fact it simply named another pattern of normal combined movements with different range and end-feel characteristics.

5 Joint "gliding" is referred to as joint "sliding" by some authors.

Gliding occurs when the same point on one joint surface comes into contact with new points on another joint surface. Pure gliding is the only movement possible between flat or congruent curved surfaces. Since there are no completely curved congruent or entirely flat joint surfaces, pure gliding does not occur in the human body.

The direction of the gliding component of joint roll-gliding associated with a particular bone rotation movement depends on whether a concave or convex articular surface is moving.

If a **concave** surface moves, joint gliding and bone movements are in the **same** direction. The moving bone and its concave joint surface are both on the same side of the axis of movement.

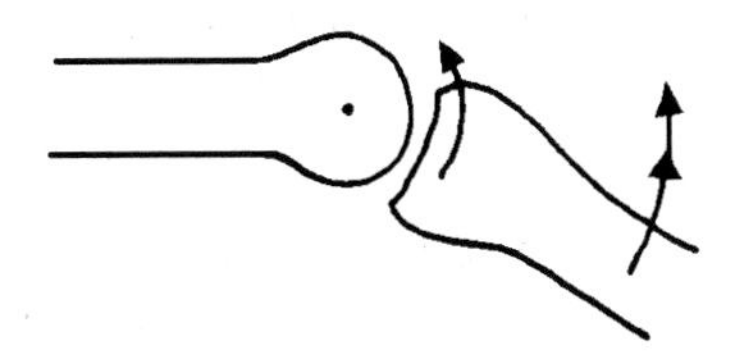

Figure 1.7
Concave surface: gliding (single arrow) in the same direction as bone movement (double arrow)

If a **convex** joint surface is moving, joint gliding and distal bone movement are in **opposite** directions. In this case, the distal aspect of the moving bone and its convex articular surface are on opposite sides of the movement axis.

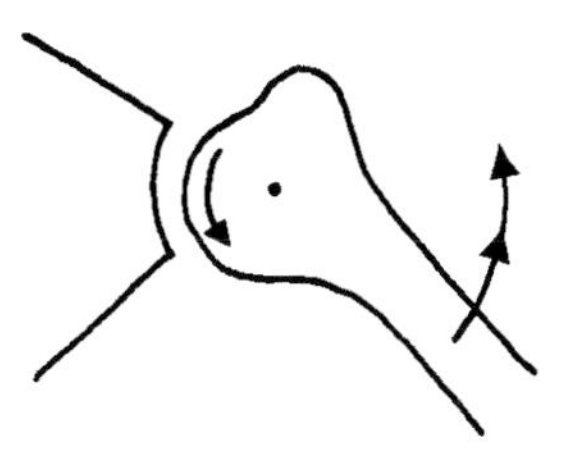

Figure 1.8
Convex surface: gliding (single arrow) in the opposite direction of the bone movement (double arrow)

Abnormal roll-gliding

With movement restrictions (hypomobility) normal joint roll-gliding is often disturbed. Usually the restricted movement is associated with an impaired gliding component which may allow joint rolling to occur without its associated gliding. Highly congruent joints, whether flat or curved, are relatively more affected by impaired gliding. A common goal in our approach to OMT is to restore the gliding component of roll-gliding to normalize movement mechanics.

Joint rolling movements in the absence of gliding can produce a damaging concentration of forces in a joint. On the same side towards which the bone is moving, joint surfaces tend to compress and pinch intraarticular structures, which can cause injury. At the same time, on the side opposite the bone movement, tissues can be overstretched.

Avoid rotational techniques for joint treatment. The following examples illustrate how damaging compression forces may occur when treating hypomobile joints with long-lever rotatoric techniques (Figure 1.9a), or with short-lever techniques applied parallel to a convex articular surface (Figure 1.9b).

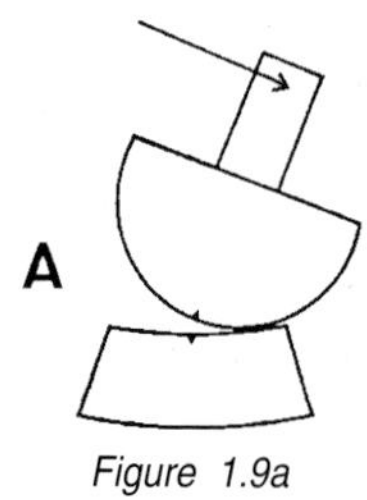

Figure 1.9a

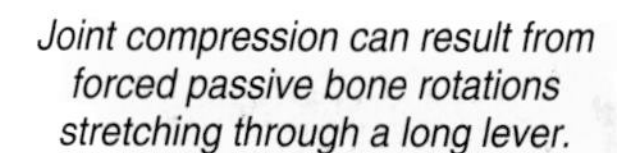

Joint compression can result from forced passive bone rotations stretching through a long lever.

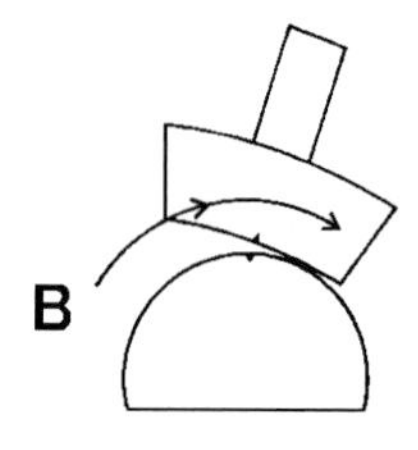

Figure 1.9b

Joint compression can result from forced passive bone rotations stretching through a short lever, or from improperly applied techniques intended to avoid compression.

If you use rotational technique for other purposes (for example, oscillations, a soft tissue stretch, or muscle stretching) be sure that the simultaneous joint gliding component occurs in an appropriate degree and direction. If you note that joint gliding is restricted or disturbed, stop the movement immediately and apply the appropriate treatment to restore joint gliding.

Remember! Joint rolling movements in the absence of gliding can produce damaging forces in a joint.

Translation of a bone

Bone translation is a linear movement of a bone along a defined axis in its respective plane. During pure translation of a bone, all parts of the bone move in a straight line, equal distances, in the same direction, and at the same speed. Bone translation can be performed only in very small increments.

Depending on the direction of the movement, bone translation can be described as parallel movement along a particular axis.

Bone translation

Longitudinal Axis Bone Translation

» **Separation** of adjacent joint surfaces, pulling them away from each other

» **Approximation** of adjacent joint surfaces, pushing them toward each other

Sagittal Axis Bone Translation

» **Ventral-Dorsal Gliding:** parallel movement of adjacent bones in relation to each other in a ventral or dorsal direction

Frontal Axis Bone Translation

» **Lateral Gliding**: parallel movement of adjacent bones in relation to each other to the right or left

In contrast to bone rotation, translation of the bone is never under voluntary control, but occurs as a consequence of external (e.g., passive movement) forces on the body.

Joint play associated with bone translation

Bone translations produce isolated traction, compression, or gliding joint play movements *in relation to the treatment plane*. These translatoric joint play movements are essential to the easy, painless performance of active movement (see *Chapter 2: Translatoric joint play*).

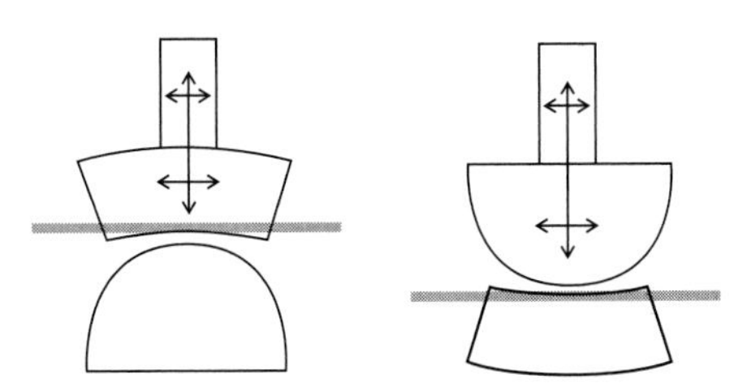

Figure 1.10
Translatoric joint play

■ Notes

2 Translatoric joint play

In every joint there are positions in which looseness or slack in the capsule and ligaments allows small, precise movements of joint play to occur as a consequence of internal and external (e.g., passive) movement forces on the body. These joint play movements are an accessory movement not under voluntary control, and are essential to the easy, painless performance of active movement.

The purpose of joint mobilization is to restore normal, painless joint function. In restricted joints, this involves the restoration of joint play to normalize the roll-gliding that is essential to active movement.

In the OMT Kaltenborn-Evjenth system we use translatoric (linear) joint play movements *in relation to the treatment plane* in both evaluation and treatment. We apply translatoric traction, compression and gliding joint play movements to evaluate joint function. We apply translatoric gliding and traction mobilizations to restore joint play.

Figure 2.1
Directions of translatoric joint play

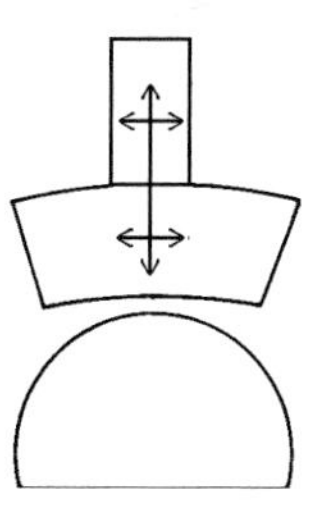

Figure 2.1a
The concave joint surface moves in relation to the stationary convex surface.

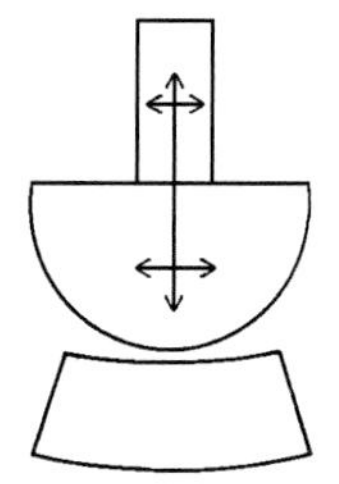

Figure 2.1b
The convex joint surface moves in relation to the stationary concave surface.

We use the term "joint play" only for translatoric (linear) movements. We do not use the term "joint play" for curved movements.

■ The Kaltenborn Treatment Plane

The Kaltenborn Treatment Plane passes through the joint and lies at a right angle to a line running from the axis of rotation in the convex bony partner, to the deepest aspect of the articulating concave surface. For practical purposes, you can quickly estimate where the treatment plane lies by imagining that it lies on the concave articular surface.

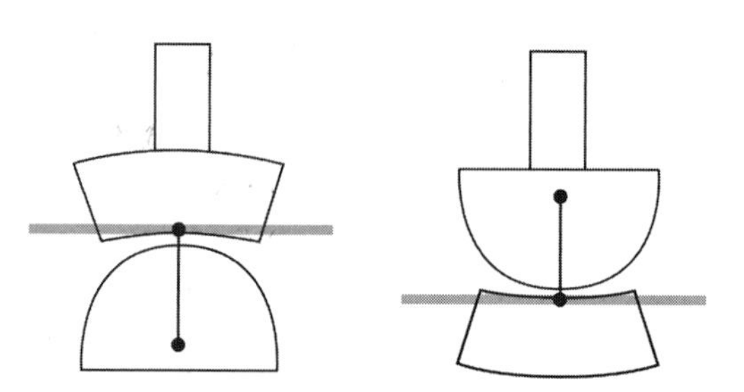

Figure 2.2
The Kaltenborn Treatment Plane lies on the concave articular surface.

The Kaltenborn Treatment Plane remains with the concave joint surface whether the moving joint partner is concave or convex.

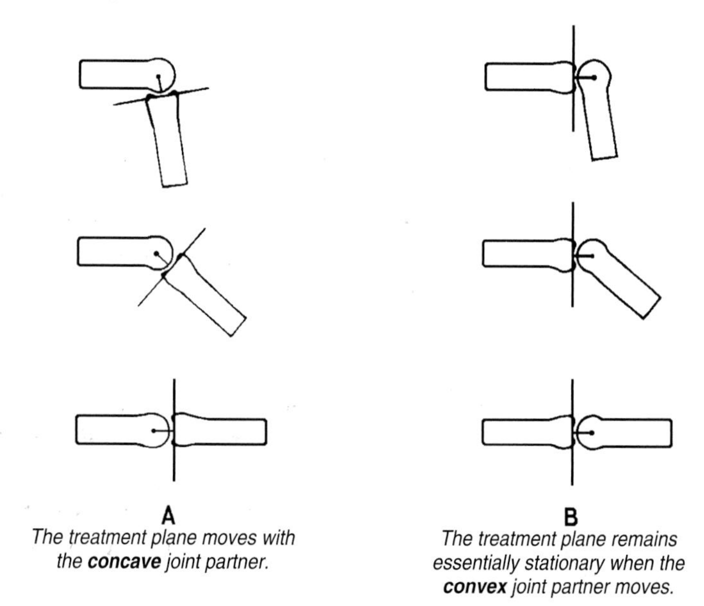

Figure 2.3
Treatment plane

Always test joint play or mobilize a joint by moving the bone parallel to, or at a right angle to, the Kaltenborn Treatment Plane.[1]

---

1 I first described the treatment plane concept in 1954 as the "joint plane" and later as the "tangential plane." The term "treatment plane" was coined by Dennis Morgan D.C., P.T. in the 1970's while collaborating with me on my writing.

■ Translatoric joint play movements

The translatoric joint play movements used in the OMT Kaltenborn-Evjenth System are traction, compression, and gliding. We define traction, compression, and gliding joint play movements in relation to the Kaltenborn Treatment Plane.

Traction

Traction (separation) is a linear translatoric joint play movement *at a right angle to and away from* the treatment plane.

Figure 2.4
Traction

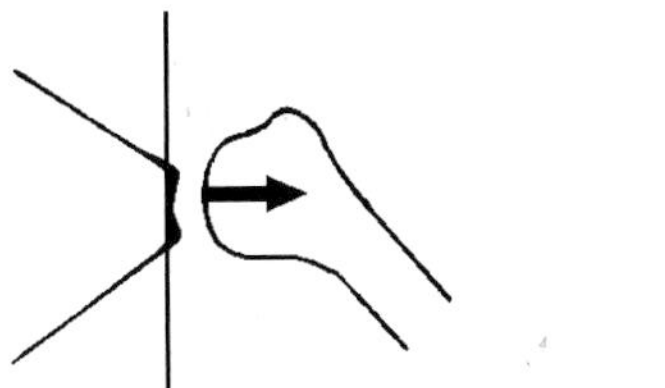

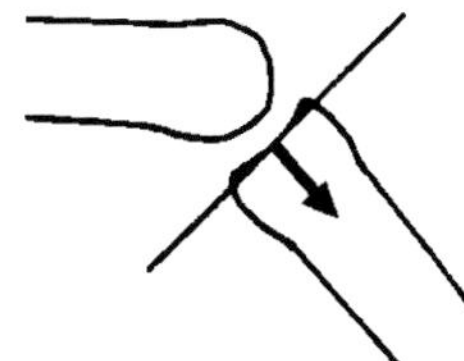

Bone movement at a right angle to and away from the treatment plane results in traction (separation) of joint surfaces.

Compression

Compression (approximation) is a linear translatoric movement *at a right angle to and toward* the treatment plane. Compression presses the joint surfaces together. Joint compression can be useful as an evaluation technique to differentiate between articular and extra-articular lesions.

Figure 2.5
Compression

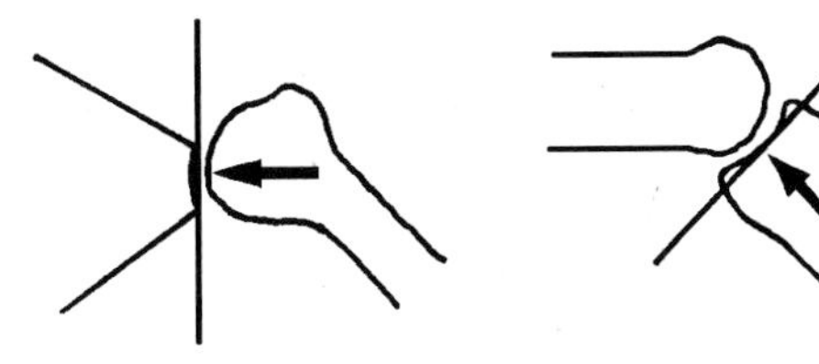

Bone movement at a right angle to and towards the treatment plane results in compression of joint surfaces.

Gliding

Translatoric gliding is a joint play movement *parallel* to the treatment plane. Translatoric gliding is possible over a short distance in all joints because curved joint surfaces are not perfectly congruent.

Grade I traction is always performed simultaneously with a translatoric gliding movement. In the figures below, the direction of gliding is indicated by two large arrows and Grade I traction by the small arrow.

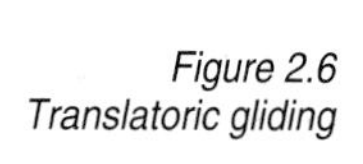

Figure 2.6
Translatoric gliding

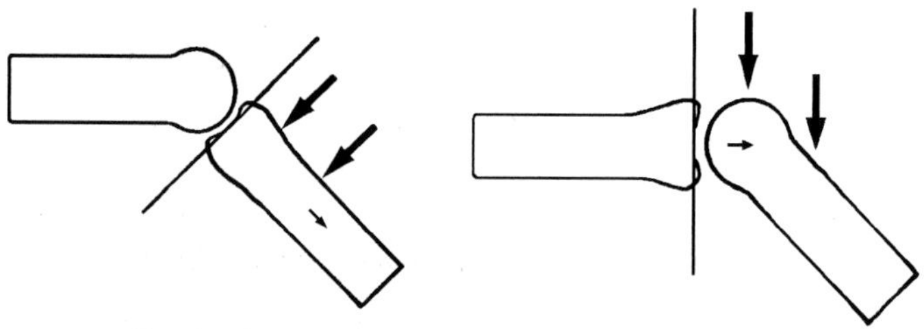

Translatoric bone movement parallel to the treatment plane resulting in translatoric gliding in the joint

Instead of using the expression "translatoric gliding," we sometimes omit the word "translatoric" or replace it with a word indicating the direction of the gliding movement. For example, we say "dorsal glide" instead of "translatoric dorsal gliding." This describes translatoric gliding of a joint in a dorsal direction as a result of passive, linear displacement of a bone.

Determining the direction of restricted gliding

There are two methods of determining the direction of restricted joint gliding: 1) the glide test, and 2) the Kaltenborn Convex-Concave Rule.

Glide test (the direct method)

Apply passive translatoric gliding movements in all possible directions and determine in which directions joint gliding is restricted. The glide test is the preferred method because it gives the most accurate information about the degree and nature of a gliding restriction, including its end-feel.

Kaltenborn Convex-Concave Rule (the indirect method)

First determine which bone rotations are decreased and whether the moving joint partner is convex or concave. Then deduce the direction of decreased joint gliding by applying the Convex-Concave Rule.

CONVEX → OPPOSITE   CONCAVE → SAME

The Kaltenborn Convex-Concave Rule is based on the relationship between normal bone rotations and the gliding component of the corresponding joint movements (roll-gliding). This approach is useful for joints with very small ranges of movement (e.g., amphiarthroses and significant hypomobility), when severe pain limits movement, or for novice practitioners not yet experienced enough to feel gliding movement with direct testing.

The most effective mobilization treatments are those that stretch shortened joint structures in the direction of the most restricted gliding. The therapist moves a bone with a *convex* joint surface *opposite* to the direction of restricted movement in the distal aspect of the bone, and a *concave* joint surface in the *same* direction as the direction of the restricted bone movement.

In both examples which follow, mobilization is in the direction of the decreased gliding component. The left joint partner is fixated (FIX) and the right partner mobilized (MOBIL). The direction of stretch, a Grade III glide mobilization, is identical to the direction of the restricted gliding component of roll-gliding.

Figure 2.7a

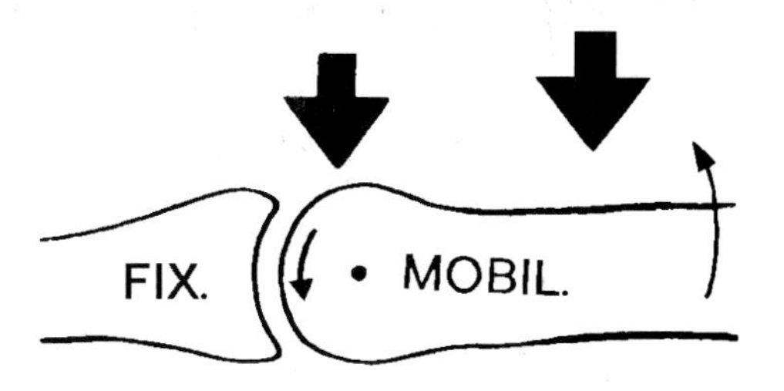

CONVEX RULE → OPPOSITE

The right (moving) joint partner's surface is convex.
When bone movement is restricted in an upward direction (curved arrow), the treatment direction is downward (two bold arrows).

Figure 2.7b

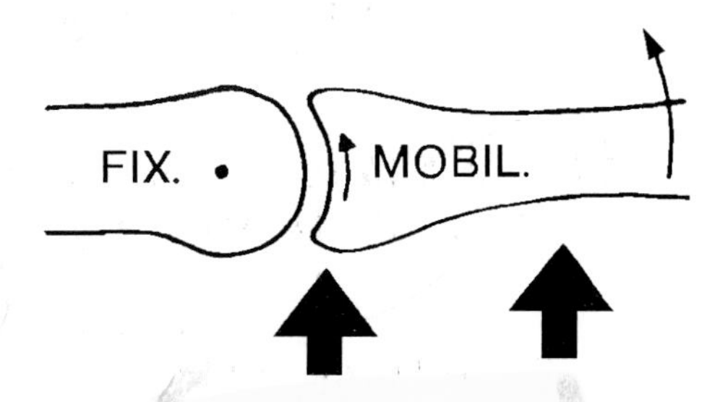

CONCAVE RULE → SAME

The right (moving) joint partner's surface is concave.
When bone movement is restricted in an upward direction (curved arrow), the treatment direction is also upwards (two bold arrows).

Grades of translatoric movement

The translatoric movements of traction and gliding are divided into three grades. These grades are determined by the amount of *joint slack* (looseness and resistance) in the joint that you feel when performing passive joint play movements.

The "slack"

The term "slack," used as a nautical expression, describes the looseness of a rope as it hangs between a boat and a dock or post. As the boat moves away from the post, the expression "taking up the slack" is used to describe tightening of the rope.

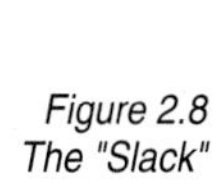

Figure 2.8
The "Slack"

All joints have a characteristic amount of joint play movement before tissues crossing the joint tighten. The amount of movement present may be of very short amplitude, but it is always present and possible to produce. This looseness or slack in the capsule and ligaments is necessary for normal joint function. The slack is taken up when testing and treating joints with gliding or traction. When gliding is performed, the slack is taken up in the direction of joint gliding; when traction is performed, the slack is taken up in the direction of traction.

The ability to correctly perform grades of movement depends on your ability to feel slackness in the joint and when tissues crossing the joint become tightened. Joint play movements are greatest, and therefore easiest to produce and palpate, in a joint's resting position where the joint capsule and ligaments are most lax.

Many factors influence the feel of joint slack being taken up, including the particular anatomy of the joint being moved, the size of the joint, the amount of soft tissues crossing the joint, the speed and smoothness of the movement, and the skill with which you perform the movement.

Normal grades of translatoric movement

I-III scale

Your ability to correctly perform translatoric movements depends on your skill in feeling when there is slack in the joint and when the tissues that cross the joint become tightened. Joint play movements are greatest, and therefore easiest to produce and palpate, in a joint's *resting position*, where the joint capsule and ligaments are most lax.

A **Grade I "loosening"** movement is an extremely small traction force which produces no appreciable increase in joint separation. Grade I traction nullifies the normal compressive forces acting on the joint.

A **Grade II "tightening"** movement first takes up the slack in the tissues surrounding the joint and then tightens the tissues. In the *Slack Zone* (SZ) at the beginning of the Grade II range there is very little resistance to passive movement. Further Grade II movement into the *Transition Zone* (TZ) tightens the tissues and the practitioner senses more resistance to passive movement. Approaching the end of the Grade II range the practitioner feels a *marked resistance*, called the *First Stop*.

A **Grade III "stretching"** movement is applied after the slack has been taken up and all tissues become taut (beyond the *Transition Zone)*. At this point, a Grade III stretching force applied over a sufficient period of time can safely stretch tissues crossing the joint. Resistance to movement increases rapidly within the Grade III range. You will find some variation in the degree of Grade III resistance among individuals and in various joints (see the dotted lines in Figure 2.9b).

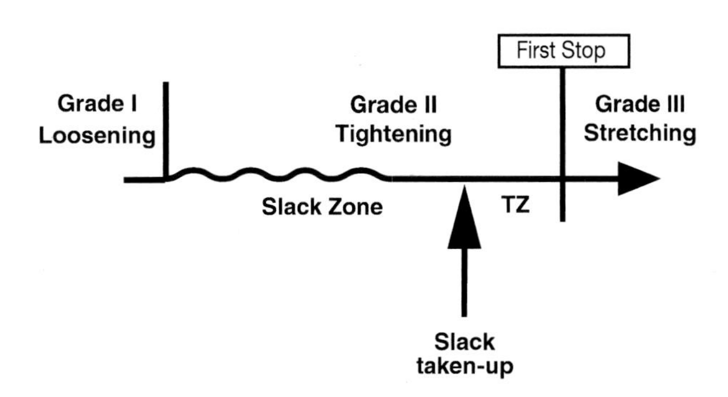

Figure 2.9a
Normal grades of movement

Palpating resistance to normal movement

In the Grade I and IISZ range the therapist senses little or no resistance. In the Grade IITZ range the therapist senses gradually increasing resistance. At the First Stop, the therapist senses marked resistance as the slack is taken up and all tissues become taut. Stretching occurs beyond this point. While in the diagrams below the slack in the Grade II translatoric movement range appears quite large, in reality it may be only millimeters long. Some practitioners apply similar grades of movement to rotatoric movements (e.g., elbow flexion), in which case the Grade II movement range could be quite large.

Figure 2.9b Relationship between resistance and grades of movement.

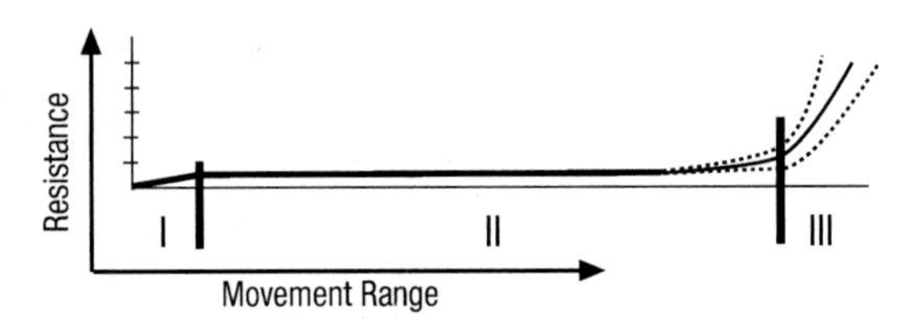

The location of the first stop can be difficult to feel. There won't be an absence of resistance suddenly followed by an abrupt stop; rather, there is a *Transition Zone*. This zone of increasing resistance may build slowly or quickly. You will feel some increasing resistance immediately before the marked resistance of the First Stop.

Mobilization for pain relief takes place in the Slack Zone and stops at the beginning of the Transition Zone, well before the marked resistance of the first stop. This is *especially important in cases of hypermobility*, since to move further could injure an undiagnosed hypermobile joint which is temporarily hypomobile ("locked") in a positional fault.

■ Pathological grades of translatoric movement

In the presence of joint pathology, the quality of end-feel is altered and grades of movement may be altered as well. For example, in the presence of a *marked hypomobility* the slack is taken up sooner than normal and greater force may be necessary to nullify intra-articular compression forces. In *hypermobility* the slack is taken up later than normal and less force may be necessary to achieve Grade I traction.

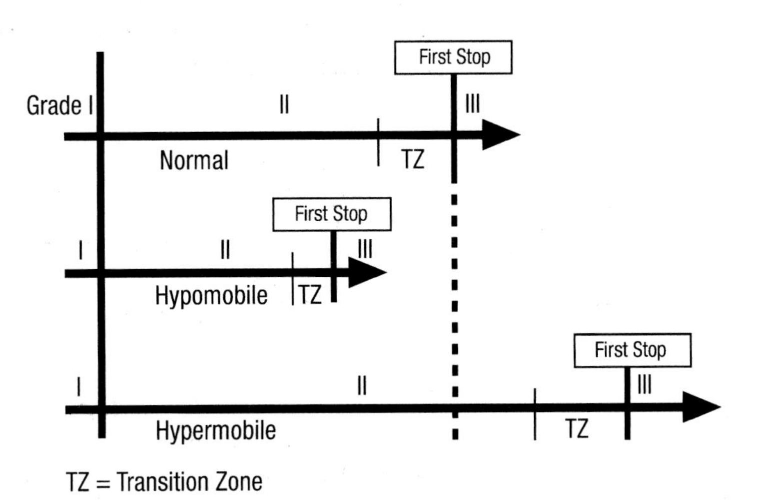

Figure 2.10a Pathological grades of movement

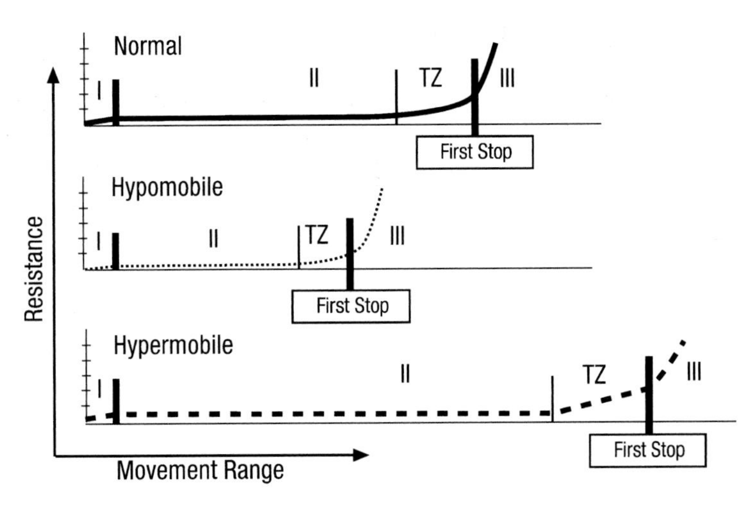

Figure 2.10b Relationship of resistance to pathological grades of movement

---

Remember: When mobilizing for pain relief, you must recognize the beginning of the Transition Zone and stop there, well before you feel the marked resistance of the First Stop.

■ Using translatoric grades of movement

Grade I

» Relieve pain with vibratory and oscillatory movements.

» Grade I traction is used simultaneously with glide tests and glide mobilizations to reduce or eliminate compression force and pain

Grade II

» Test joint play traction and glide movements.

» Relieve pain. (Treatment takes place in the Slack Zone, not in the Transition Zone.)

» Increase or maintain movement, for example when pain or muscle spasm limits movement in the absence of shortened tissue. (Relaxation mobilization can be applied within the entire Grade II range, including the Transition Zone.)

Grade III

» Test joint play end-feel.

» Increase mobility and joint play by stretching shortened tissues.

3 Tests of function

A test of function enables you to *see*, *hear*, and *feel* the patient's complaints. The constellation of symptoms and signs that emerges from tests of function differentiate the nature of the structures involved in the dysfunction, for example, whether these are muscles or joints, and allows you to apply treatment specifically to those structures. Tests of function are a key element within the OMT evaluation (see *Chapter 4: OMT evaluation).*

Tests of function

A. Active and passive rotatoric (angular) movements: Identify location, type, and severity of dysfunction.

- *Standard (Anatomical, Uniaxial) movements*
- *Combined (Functional, Multiaxial) movements*

B. Translatoric joint play movements: Further differentiate articular from nonarticular lesions; identify directions of joint restrictions.

- *Traction*
- *Compression*
- *Gliding*

C. Resisted movements: Test neuromuscular integrity and status of associated joints, nerves and vascular supply.

D. Passive soft tissue movements: Differentiate joint from soft tissue dysfunction and the type of soft tissue involvement.

- *Physiological movements*
- *Accessory movements*

E. Additional tests

■ Principles of function testing

Be specific when asking the patient about symptoms during the examination. Ask the patient to describe the character and distribution of their symptoms or if already existing symptoms change with each test procedure. Especially note if a particular movement provokes the primary complaint for which the patient seeks treatment.

■ Assessing quantity of movement

Examine the range of motion for each standard and combined movement first by observing the active movement. Then continue the same movement passively with overpressure. The passive part of the movement is *not* started again at the zero position, but begins where the active movement stops. In this way you can compare the range of active movement with the same passive movement.

The results of this test may reveal *hypomobility,* defined as movement less than established norms, or *hypermobility*, defined as movement greater than established norms. Note also that a joint can be hypomobile in one direction and hypermobile in another.

Hypomobility or hypermobility are only pathological findings if they are associated with symptoms (for example, positive symptom provocation or alleviation tests) and if the associated end-feel is pathological. Hypomobility or hypermobility with a normal end-feel is usually due to a congenital structural anomaly or a normal anatomic variation and is unlikely to be symptomatic or to benefit from mobilization treatment. Remember that movement quantity tests alone cannot differentiate the nature of the dysfunction, but can implicate a capsular pattern or significant muscle shortening.

With larger passive movements, test range of movement slowly through an entire range to the first significant stop. With smaller passive movements in joints with little range of movement, test range of movement first with more rapid oscillatory movements that do not require stabilization of neighboring joints. If these oscillatory tests reveal restrictions or symptomatic areas, follow up with more careful evaluation of the movement range using slower movements and stabilization of the adjacent joints.

The amount of active or passive joint movement can be measured with an instrument such as a goniometer, ruler, or other device (e.g., distance of fingertips to floor as a measurement of standard rotatoric spine and hip movement). Measure standard bone movements from the zero position around their defined axes.

Hypomobility or hypermobility are only pathological findings if they are associated with symptoms and a pathological end-feel.

Manual grading of rotatoric movement (0-to-6 scale)[1]

In joints with little range of motion such as the carpal joints or single spinal segments, it may be impossible or impractical to measure range of motion with a goniometer. Range of motion may then be tested manually and classified using the following scale:

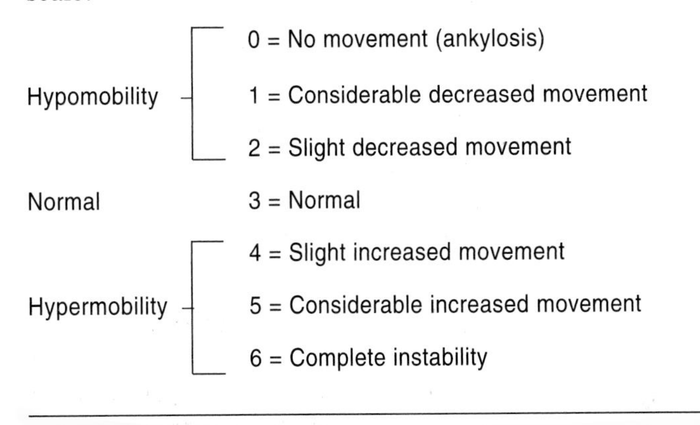

A joint can be both hypomobile in one direction and hypermobile in another.

■ Assessing quality of movement

The ability to *see* and *feel* movement quality is of special significance, as slight alterations from normal may often be the only clue to a correct diagnosis.

Assess movements with minimal forces so as not to obscure slight deviations from normal. Repeat each passive movement at different speeds to reveal various types of restrictions. For example, slower passive movements are more likely to reveal joint restrictions, while more rapid movements can trigger abnormal muscle reactivity.

Passive movement quality is best assessed throughout an entire range of movement to the first significant stop. Important findings are easily overlooked if passive movement is tested only at the limit of active movement (overpressure), since the first significant finding in a passive movement may be detected before the first stop.

---

1 The 0-to-6 scale for manual grading of rotatoric movement was originally based on Stoddard's 1-to-4 scale and was later revised and expanded by Paris. Paris's expanded concept was first presented at the 1977 IFOMT meeting (Vail, CO) as part of the Kaltenborn-Evjenth sessions.

Quality of movement to the first stop

Test movement quality by first observing the active movement, then feel the same movement passively until you meet the first significant resistance. Apply minimal force and perform the movement slowly several times throughout the entire range of motion.

Note quality of movement from the very beginning of the range of movement up to the first stop. Passive movements should be free, smooth, and independent of the speed with which they are carried out. Deviations from normal can often be detected as soon as you contact the patient or very early in the range of movement. Be alert to slight abnormalities from the very instant you contact the patient.

End-feel: Quality of movement after the first stop

End-feel is the sensation imparted to your hands at the limit of the available range of movement. Test end-feel with a slight additional stretch after the first significant stop of a *passive* movement (quality test). Note that end-feel testing is not the same as overpressure applied after an *active* movement (quantity test).

End-feel can be evaluated during standard and combined passive rotatoric movements (overpressure end-feel) or during translatoric joint play movements (joint play end-feel).

Figure 3.1
End-feel

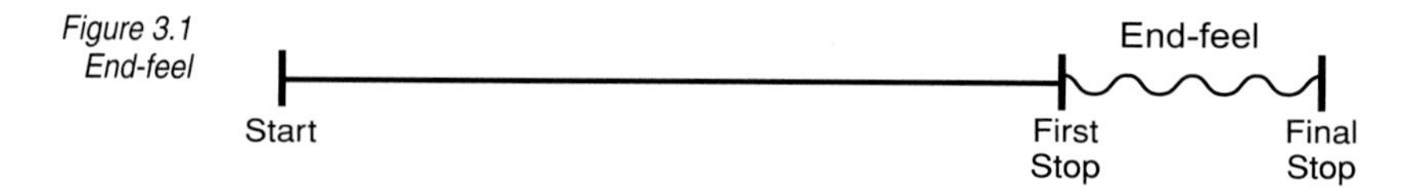

Evaluate end-feel slowly and carefully after a passive movement from the zero position (or actual resting position) through the entire range of movement past the first stop (a slight additional stretch) to the final stop. Subtle end-feel findings are easily overlooked if you test end-feel too quickly, or if you test an insufficient range of movement.

Normal physiological end-feel

Each joint movement has a characteristic end-feel, depending on the anatomy of the joint and the direction of movement tested. End-feel also varies with each individual, depending on age, body type and build. After the first significant resistance to passive movement is met (first stop), carefully apply a small additional stretch to determine whether the end-feel is soft, firm, or hard.

» **Soft:** A soft end-feel is characteristic of soft tissue approximation (e.g., knee flexion) or soft tissue stretching (e.g., ankle dorsiflexion).

» **Firm:** A firm end-feel is characteristic of capsular or ligamentous stretching (e.g., medial or lateral rotation of the humerus and femur). A normal capsular end-feel is less firm (firm "-") and a normal ligamentous end-feel is more firm (firm "+"). A firm end-feel is variable among individuals depending on many factors, including the size and age of the individual and the extent of degenerative changes.

» **Hard**: A hard end-feel occurs when bone or cartilage meet (e.g., elbow extension and flexion).

All three types of normal joint end-feels have an *elastic* quality to varying degrees. When overpressure is released, the joint rebounds back to the first stop or further into the slack (Grade II range). Normal end-feels are pain free.

Remember: Normal end-feels are pain free.

It is important for a manual therapist to be able to differentiate joint from muscle end-feels, and normal (physiological) from pathological end-feels.

Pathological end-feel

A pathological end-feel is one that occurs at *another place* and is of *another quality* than is characteristic for the joint being tested. In other words, the stop may be met earlier or later in the range of motion than is normal, and the quality of the stop is uncharacteristic for the joint being examined. For example, scar tissue imparts a firmer, less elastic end-feel; muscle spasm produces a more elastic and less soft end-feel; shortened connective tissue (for example, fascia, capsules, ligaments) gives a firmer, less elastic end-feel; intra-articular swelling produces a soft resistance just before or instead of the movement's usual end-feel *(boggy end-feel)*. With hypermobility or ligamentous laxity, you will find a final stop later in the movement range and with a softer end-feel than normal. Some end-feels are characteristic of specific pathologies and are usually tested with rotation bone movements. For example, a displaced meniscus can impart a *springy end-feel.*

A pathological end-feel is judged to be less elastic if the movement does not rebound back to its first stop when testing pressure is released.

The patient may guard against end-feel testing or ask that the movement be discontinued before you reach their "true" end-feel. This is called an *empty end-feel.* The empty end-feel is a response to severe pain or muscle spasm secondary to conditions such as fractures or acute inflammatory processes, or can be psychogenic in origin.

It is possible for the same joint to present with a normal end-feel in one movement direction and a pathological end-feel in another direction. Indications and contraindications for treatment based on end-feel findings only apply to the impaired movement direction. For example, a hard, inelastic end-feel only contraindicates Grade III stretch mobilization in the direction that is restricted.

Pathological end-feel findings can be subtle and may be apparent only to the most skilled practitioner. A symptomatic joint may appear to have normal range of movement to the novice, while the experienced practitioner will discover an abnormal end-feel. A novice practitioner usually needs an immediate and careful comparison with a normal joint to recognize the pathological character of an end-feel. By the same token, a novice practitioner may inappropriately judge that a joint with less than expected range of movement requires treatment, while the skilled practitioner would assess a normal end-feel with no associated muscle reactivity, and would judge the joint as normal.

■ Elements of function testing

■ Active and passive rotatoric movements

Active **movements** quickly provide a general indication of the location and type of dysfunction as well as its severity. Active movements require patient cooperation, upper and lower motor neuron integrity, and normal muscle and joint function. Since active movements stress both joints and soft tissue, any positive findings can only be interpreted in light of additional tests of function, particularly passive movement testing.

Passive movements provide additional information by allowing you to *feel the quality of movement and end-feel.* Sensing the type of resistance through the entire range of movement, including how the movement stops, provides valuable diagnostic information. Slight alterations from normal may be the only clue to a diagnosis.

Passive movements are normally greater than the corresponding active movements.

It is possible to differentiate between lesions involving contractile or non-contractile elements by comparing responses to various types of passive movements. For example, careful examination of passive movements allows you to detect muscle shortening, a capsular pattern, hypomobility or hypermobility. As with all examination procedures, note if there is any production or alteration of symptoms. Compare the results with accepted norms or with the same movement in the opposite joint.

There are two general categories of active and passive rotatoric joint movements which are used for different purposes in an OMT evaluation:

» **Standard (anatomical) movements,** e.g., flexion, extension, abduction, adduction and rotation, occur in the cardinal planes and around defined axes. They are used for measurement and to reveal asymmetries and disturbances in movement quality (for example, a painful arc). Since these movements are standard and generally recognized, they facilitate communication between therapists and physicians.

» **Combined (functional) movements,** e.g., coupled and noncoupled movements, occur around multiple axes and in multiple planes and allow you to specifically stress various tissues and structures. These movements are useful in understanding and analyzing the exact mechanism of injury and reproducing the patient's chief complaint. It is not unusual to perform combined movements in order to reveal subtle lesions that could be overlooked with standard movement testing alone.

Changes in the quantity and quality of rotatoric movement can be due to lesions within the joint or the surrounding soft tissue and may manifest themselves in the form of a painful arc, capsular pattern, or muscle shortening. Specific rotatoric bone movement is also used to test neural tension and mobility.

Painful arc

Pain occurring anywhere in the range of active and/or passive movement which is preceded and followed by no pain is called a painful arc, according to Cyriax. A painful arc implies that a pain-sensitive tissue is being squeezed between hard structures. Deviations from the normal path of movement may be an attempt by the patient to avoid such pain. It is important to note such deviations in order to not overlook a painful arc.

Capsular pattern

If the entire capsule is shortened, we find what Cyriax calls a capsular pattern. The capsular pattern manifests itself as a characteristic pattern of decreased movements at a joint. When expressing the capsular pattern, a series of three or four movements are listed in sequence: the first movement listed is proportionally most decreased, the second movement listed is next decreased, and so on.

We describe typical capsular patterns in each joint's respective chapter. For example: shoulder = external rotation-abduction-internal rotation; this denotes that external rotation is proportionally the most decreased movement, abduction the next decreased, and internal rotation the least decreased movement.

A capsular pattern is usually present when the entire capsule is affected (e.g., inflammatory arthritic conditions). However, limitation of movement due to capsular shortening does not necessarily follow a typical pattern. For example, only one part of a capsule may be shortened due to trauma, surgery, inactivity, or some other localized lesion of the capsule. In these cases, limitation of movement will be evident only with movements that stretch the affected part of the capsule.

Testing rotatoric movements

During active movements, observe the patient's range and quality of movement and at the same time note any crepitus or change in the patient's symptoms. Ask the patient to describe symptoms or abnormal sensations such as a painful arc. Repeat active joint movements several times while you observe from the back, the front, and the sides.

Be specific when asking the patient about symptoms during examination. Ask the patient to describe the character and distribution of pain or if already existing symptoms change with each test procedure. Especially note if a particular movement provokes the same pain the patient complains of during daily activities.

Observe whether a movement is smooth and if there is angularity or asymmetry, or change in the patient's symptoms or abnormal sensations, such as a painful arc.

When possible, continue the movement achieved actively with gentle passive overpressure, moving the joint to the last stop while the patient relaxes (quantity test). Note that this is not an evaluation of end-feel (quality test), but a way to determine whether a joint lesion is limiting the active movement. Range of movement with passive overpressure is normally greater than the corresponding active movement. If passive overpressure produces little or no increase in the active movement range, the movement is probably limited by a joint structure.

Passive range of movement with overpressure is normally greater than the corresponding active movement.

Examine passive bone rotations as general movements and as specific movements.

Differentiating articular and extra-articular dysfunction

Cyriax provides one model for distinguishing contractile (muscle) lesions from noncontractile (e.g., joint) lesions by comparing responses to various tests of active and passive movement. Cyriax divides musculoskeletal structures into contractile and noncontractile elements for diagnostic purposes. The contractile elements consist of the muscle with its tendons and attachments. Noncontractile elements include all other structures such as bones, joint capsules, ligaments, bursae, fasciae, dura mater, and nerve roots.

Noncontractile Dysfunction

» Active and passive movements produce or increase symptoms and are restricted in the same direction and at the same point in the range.

Example: Active and passive external rotation of the shoulder is painful and/or restricted at the same degree of range.

» Passive joint play movements produce or increase symptoms and are restricted.

» Resisted movements are symptom free.

Contractile Dysfunction

» Active and passive movements produce or increase symptoms and are restricted in opposite directions.

Example: Active external rotation of the shoulder is painful and restricted as the affected muscle contracts; passive external rotation is pain free and shows a greater range of movement; passive internal rotation is painful as the affected muscle is stretched.

» Passive joint play movements are normal and symptom free.

» Resisted movements produce or increase symptoms.

While Cyriax's differentiation process produces clear findings in many musculoskeletal lesions of the extremities, interpretation of findings can be less clear in some pathologies, such as with the presence of subtle contractile tissue lesions, in cases where a significant inflammatory process produces pain during a resisted tests, or when the muscle contraction produce symptomatic joint compression in underlying dysfunctional joints. Therefore, joint dysfunction must be confirmed with joint testing first. For example, traction-alleviation and compression-provocation tests may reveal joint dysfunction.

If you determine that a joint structure is involved, focus the OMT evaluation on more specifically identifying the nature and location of the joint dysfunction so that you can select a more specific, and thus more effective, treatment approach.

Differentiating muscle shortening from muscle spasm

A skilled practitioner can usually tell the difference between muscle connective tissue shortening and muscle spasm based on end-feel testing. A shortened, tight muscle imparts a firmer, less elastic end-feel, while muscle spasm produces a more elastic and less soft end-feel, sometimes accompanied by increased muscle reactivity.

Novice practitioners may make the same differentiation based on the patient's response to a specific muscle relaxation maneuver. For example, in the case where a patient's hamstrings limit a straight-leg-raise movement, the practitioner positions the limb at the limit of available motion, and then performs a "hold-relax" muscle relaxation maneuver on the hamstrings. In the relaxation period immediately following the muscle contraction, a muscle in spasm will relax sufficiently to allow some elongation of the muscle and the straight-leg-raise range will increase. A shortened muscle will not allow increased movement into the range without additional sustained stretching.

■ Translatoric joint play tests

Testing the quantity and quality of joint play, including end-feel, is always a part of the examination of extremity joints.

Evaluate joint play using traction, compression, and gliding in all of the translatoric directions in which a joint is capable of moving.

Joint play range of movement is greatest in the resting position of the joint and therefore easiest to feel in this position. The practitioner with advanced skill also evaluates joint play outside the resting position, where a naturally smaller range of movement can make the movement more challenging to palpate.

There are two ways to test joint play:

1) Fixate one joint partner and move the other through the fullest possible range of joint play movement. Feel for changes in the resistance to the movement through Grade II, past the first stop, and into Grade III for end-feel. Determine whether there is normal movement quality through the range and if there is hypo- or hypermobility.

2) Apply vibrations, oscillations, or small amplitude joint play movements while you palpate the joint space. Apply no fixation or stabilization. This method of joint play testing is especially useful for spinal joint testing.

Ask the patient if there are symptoms during movement and note if pain affects the quality or quantity of movement.

Traction and compression tests

Since traction often relieves and compression often aggravates joint pain, these joint play movements help determine if an articular lesion exists. Resisted movements produce some joint compression, so it is important to test joint compression separately and before resisted tests.

It is important to test joint compression separately and before resisted tests, since resisted movements also produce joint compression.

If the patient has symptoms with **traction tests** in the normal resting position, use three-dimensional positioning to find a position of greater comfort (i.e., the actual resting position) and reevaluate the patient's response to traction.

If a general **compression test** produces the patient's complaints, you may need to limit further evaluative techniques that cause joint compression, for example, resistive tests or other techniques that produce secondary joint compression forces.

If compression tests in the resting position are negative, and if no other tests of function provoke or increase the patient's complaint, compression tests should also be performed in various three-dimensional positions. In some subtle joint dysfunctions, this may be the only way to locate a patient's lesion.

Gliding tests

Examination of translatoric gliding movements helps further differentiate articular from extra-articular lesions, since gliding primarily tests those structures belonging to the anatomical joint. Gliding movements are also important for determining the specific directions of joint movement restrictions. The skilled manual therapist evaluates gliding movement both in the joint's resting position and in various positions outside the resting position.

■ Resisted movements

Resisted tests simultaneously evaluate neuromuscular integrity, the contractile elements, and, indirectly, the status of associated joints, nerves, and vascular supplies.

According to Cyriax, a resisted test must elicit a maximal muscle contraction while the joint is held still near its mid-position (resting position). Not allowing movement during a resisted test will help eliminate the joint as the source of pain; however, a certain amount of joint compression and gliding is inevitable. To exclude pain arising as a result of joint dysfunction, compression tests should be performed *before* the resisted test. Therefore, if compression tests provoke pain, resisted tests are of limited value. Cyriax interprets resisted tests in the following ways:

Painful and strong	=	minor lesion of a muscle or tendon
Painful and weak	=	major lesion of a muscle or tendon
Painless and weak	=	neurological lesion or complete rupture of a muscle or tendon
Painless and strong	=	normal

There are three general methods of performing resisted tests: manual muscle testing (standard positions and methods); machines (for example, tensiometers and various isokinetic testing devices); and specific functional maneuvers (for example, proprioceptive neuromuscular facilitation techniques).

When testing large muscle performance with manual resistance, the potentially strong muscle contractions are best controlled if the therapist induces the force. The patient should resist your attempt to move them ("hold") in response to your instruction, "Don't let me move you." The patient should not try to push or pull against you, nor should you instruct them to "Push" or "Pull."

Differential diagnosis for pain in a muscle synergy

Several muscles usually act together in a synergy to perform a particular movement. All muscles which normally function in a synergy contract regardless of joint position. Electromyographic studies have shown that muscle activity is not significantly affected by changes in joint position. Therefore, conventional manual strength testing performed from various positions cannot reliably differentiate the source of a musculotendinous pain.

To identify a specific muscle or tendon responsible for a patient's pain, the examiner selectively elicits or prevents contraction of a specific muscle or group of muscles. There are three methods described below.

Testing a muscle's secondary function in the same joint

If one muscle in a joint movement synergy has a secondary function not shared by the other muscles in the synergy, it can be selectively tested. For example, if resisted knee flexion is painful, further examination of resisted lateral and medial leg rotation may identify the specific muscle causing the pain. If lateral rotation is painful and medial rotation is not, then it is likely that the biceps femoris is injured and not the other knee flexors which medially rotate the leg.

In the chapters on joint techniques, the secondary functions of relevant muscles are included in the examination schemes as "Other functions."

Testing a muscle's secondary function at an adjacent joint

A muscle or tendon can be selectively stressed if it is the only muscle in a synergy which functions at another joint. For example, pain with resisted shoulder flexion can be due to a lesion in one of several muscles in a synergy producing this movement. If resisted elbow flexion produces the same pain, then the biceps is implicated as it is the only muscle which can produce both shoulder and elbow flexion.

Testing using reciprocal inhibition

Selectively relaxing a muscle may be useful as a differential diagnostic procedure. This technique uses the concept of "reciprocal inhibition" to prevent a muscle from contracting in synergy with other muscles during a movement. This is accomplished by resisting the antagonist of the muscle to be eliminated at the same time as the test movement is resisted.

An example illustrating this procedure is the differentiation between muscles extending the wrist and those extending the fingers. To eliminate the wrist extensors and test the finger extensors, the examiner resists palmar flexion at the wrist and finger extension simultaneously; the resisted wrist palmar flexion will inhibit contraction of the wrist extensors. To eliminate the finger extensors and test the wrist extensors, finger flexion and wrist extension are resisted simultaneously; a reflex relaxation of the finger extensors is accomplished by resisting finger flexion.

Passive soft tissue movements

Soft tissues are examined similarly to joints, using passive movements to assess the quantity and quality of movement and pain. There are two major types of passive soft tissue movements: physiological and accessory movements.

Physiological movements (muscle length and end-feel)

Test soft tissue length and end-feel by moving a limb or bone so that muscle attachments are moved maximally apart (lengthened). It is often necessary to use combined movements to achieve full tissue lengthening. Soft tissue end-feel testing during lengthening is particularly important to help differentiate joint from soft tissue dysfunction and to determine the type of soft tissue dysfunction. For example, muscle spasm will have a less firm end-feel than a muscle contracture.

It is not unusual for joint structures to limit movement before a position of muscle stretch can be attained, especially in the presence of chronic joint disorders with associated degenerative changes. Muscle length testing requires that you be knowledgeable about muscle functions, muscle attachments, and muscle relationships to each joint they cross. These techniques are thoroughly described in the textbooks *Muscle Stretching in Manual Therapy, Vol. I and Vol. II* and *Autostretching* by Olaf Evjenth and Jern Hamberg.

Accessory soft tissue movements

Examination of accessory soft tissue movement tests the elasticity, mobility, and texture of soft tissues. Accessory soft tissue movement cannot be performed actively, but is tested by passively manipulating soft tissues in all directions. Skillful technique can help pinpoint localized changes in soft tissue texture due to, for example, scar tissue, edema, adhesions, and muscle spasm.

Muscle play is an accessory soft tissue movement. Muscle play testing involves manually moving muscles in transverse, oblique, and parallel directions in relation to the muscle fibers. A passive lateral movement of muscle is one example of muscle play.

■ Additional tests

Additional examination procedures may be necessary, including assessment of coordination, speed, endurance, functional work capacity, and work site ergonomic evaluations. These exams do not always have to be complicated, expensive, or require special equipment in order to give valid, useful and important information.

4 OMT evaluation

Goals of the OMT evaluation

The OMT evaluation is directed toward three goals:

1) **Physical diagnosis**
To establish a physical, or biomechanical, diagnosis.

2) **Indications and contraindications**
To identify indications and contraindications to treatment.

3) **Measuring progress**
To establish a baseline for measuring progress.

The emphasis of an OMT evaluation varies depending on the purpose of the patient visit and the setting in which the manual therapist practices.

Physical diagnosis

The skilled manual therapist can *hear* (via the patient history) and *see* and *feel* (via the physical exam) a patient's physical diagnosis.

The physical diagnosis is based on a model of somatic dysfunction that assumes a highly interdependent relationship between musculoskeletal symptoms and signs.[1] In the presence of somatic dysfunction there is a correlation between the patient's musculoskeletal signs and the production, increase, or alleviation of symptoms during a relevant examination procedure.

Musculoskeletal conditions that respond well to treatment by manual therapy typically present with a clear relationship between signs and symptoms. An OMT evaluation that shows no correlation between signs and symptoms usually indicates that the patient's problem originates from outside of the musculoskeletal system and that mechanical forms of treatment such as manual therapy are less likely to help.

1 The concept of somatic dysfunction was originally used by osteopaths to better describe and reflect the many somatic interrelated aspects of a musculoskeletal disorder.

Common characteristics of somatic dysfunction

Symptoms (history)

- pain, weakness, stiffness, numbness, headache, dizziness, nausea, etc.

Signs (physical examination findings)

A. Soft tissue changes
   - altered tissue tension, elasticity, shape, texture, color, temperature, etc.

B. Functional changes
   - impaired strength, endurance, coordination
   - impaired mobility:
     - joints (e.g., hypomobility or hypermobility)
     - soft tissues (e.g., contractures)
     - neural and vascular elements (e.g., entrapment syndromes, neural tension signs)

OMT examination techniques are designed to reveal the subjective and objective manifestations of somatic dysfunction. You must be able to distinguish between them in order to administer appropriate treatment. For example, a patient's inability to straighten the knee may be due to pain or soft tissue contractures, peripheral neuropathy, intra-articular swelling, primary muscle disease, lumbar radiculopathy, or a meniscal block.

The OMT practitioner emphasizes three major differential diagnostic decisions in the evaluation of a somatic dysfunction:

1. Determining whether a problem is primarily in the anatomical joint or associated soft tissues, including neural structures (e.g., the "physiological joint")
2. Deciding if joint hypomobility or hypermobility is present
3. Determining whether treatment should be directed toward pain control or biomechanical dysfunction

A manual therapist skilled in mobility testing can often palpate a somatic dysfunction before it can be medically diagnosed. For example, symptoms of nontraumatic origin (usually pain) associated with arthroses, discopathies, or segmental pain syndromes with radiating pain are often associated with a palpable alteration in movement quality (e.g., an abnormal end-feel). In the early stages of pathology, this subtle alteration in movement quality may be palpable long before there is restriction in range of movement and before the pathology is apparent on diagnostic imaging studies.

The role of the manual therapist in making a **physical diagnosis** varies in different practice settings. Most often a referring physician establishes a **medical diagnosis** that implicates the musculoskeletal system and rules out serious pathology that might mimic a musculoskeletal disorder. In this case, the manual therapist typically omits the organ system review and family history from the OMT evaluation. Emphasis is on the more detailed biomechanical and functional assessment necessary to identify the structures involved (refinement of the medical diagnosis) and the functional status of their involvement (the physical diagnosis).

The manual therapist confirms the initial physical diagnosis of somatic dysfunction with a low-risk **trial treatment** as an additional evaluation procedure. For example, traction is the most common trial treatment for a joint hypomobility. If the trial treatment does not alleviate symptoms or if symptoms are worsened, further evaluation is necessary and a different trial treatment is tested.

The physical diagnosis is further refined through ongoing assessments of each subsequent treatment. The results of these reassessments are an ongoing part of the evaluation process.

Indications and contraindications

No treatment performed on a living subject is guaranteed to be free of risk or complications. Conscientious patient evaluation and appropriate selection of techniques minimize the potential risks of manual treatment.

Indications

Indications for treatment by manual therapy are based more on the physical diagnosis than on the medical diagnosis.

Restricted joint play (hypomobility) and an abnormal end-feel are the two most important criteria for deciding if mobilization is indicated. Grade III stretch mobilization is indicated when a movement restriction (hypomobility) has an abnormal end-feel and appears related to the patient's symptoms. Hypomobility presenting with a normal end-feel and no symptoms is not considered pathological, and is not treated. In such cases, the movement restriction is either due to a congenital anatomical variation, or the symptoms in that area are referred from another structure.

In patients who cannot yet tolerate examination or specific treatment with a biomechanically significant force, within-the-slack (Grades I-IISZ) mobilizations and other palliative modalities provide short-term symptom relief. These symptom control treatments are primarily used as a temporary measure to prepare a patient to tolerate further specific examination or more intensive treatments (for example, a Grade III stretch movement) that will produce a more lasting effect.

In patients with hypomobility due to muscle spasm in the absence of tissue shortening, relaxation mobilizations in the Grade I - II range are generally effective.

In the presence of excessive joint play (hypermobility), stabilizing (limiting) measures are indicated and Grade III stretch mobilization is contraindicated.

Contraindications

Contraindications to manual therapy are relative and depend on many factors, including the vigor of the technique, the medical and physical diagnoses, the stage of pathology, the relationship between specific musculoskeletal findings such as joint play range of movement and joint play end-feel, and the patient's symptoms. In other circumstances good professional judgment limits the use of any manual contact technique, for example, in the case of patient resistance to treatment or unwillingness to cooperate.

Grade I and II "within-the-slack" mobilizations are seldom contraindicated, but many contraindications exist for Grade III stretch mobilizations. There are additional specific contraindications for Grade III manipulative (high velocity thrust) techniques which are performed so quickly that the patient is unable to abort the procedure. Thrust procedures require a high level of skill and knowledge to apply safely and are not covered in this basic book.

General contraindications to Grade III stretch mobilization relate primarily to health problems that reduce the body's tolerance to mechanical forces and therefore increase the risk of injury from stretch mobilization treatment. For example:

- » pathological changes due to neoplasm, inflammation, infections, or osteopenia (e.g., osteoporosis, osteomalacia)
- » active collagen vascular disorders
- » massive degenerative changes
- » loss of skeletal or ligamentous stability in the spine (e.g., secondary to inflammation or infection or after trauma)
- » certain congenital anomalies
- » anomalies or pathological changes in vessels
- » coagulation problems (e.g., anticoagulation factors, hemophilia)
- » dermatological problems aggravated by skin contact and open or healing skin lesions

Grade III stretch mobilization is contraindicated for joints with active inflammation. However, the presence of a progressive inflammatory disease, such as rheumatic disease is not an absolute contraindication for Grade III stretch mobilization. During a quiescent stage of illness when the joint involved is not inflamed, it can often be safely stretched beyond its slack.

Mobilization may also be contraindicated in certain autonomic nervous system disorders because mobilization can affect autonomic responses. For example, in patients with autonomic disturbances associated with diabetes mellitus there have been reported cases of thoracic mobilization triggering hyperventilation, low sugar levels, or loss of consciousness.

Specific contraindications to Grade III stretch mobilization techniques include:

» decreased joint play with a hard, nonelastic end-feel in a hypomobile movement direction

» increased joint play with a very soft, elastic end-feel in a hypermobile movement direction

» pain and protective muscle spasm during mobilization

» positive screening tests

Screening tests identify conditions that contraindicate specific mobilization techniques and should be completed prior to treatment.

■ Measuring progress

Changes in a patient's condition are assessed by monitoring changes in one or more dominant symptom and comparing these changes with routine screening tests and the patient's dominant signs.

A relevant sign is one that is reproducible and related to the patient's chief complaints. That is, the sign improves as the patient's symptoms improve, and the sign worsens as the patient's symptoms worsen.

Periodic reassessment of the patient's chief complaints and dominant physical signs during a treatment session guides treatment progression. If reassessment reveals normalization of function (e.g., mobility) along with decreased symptoms, then treatment may continue as before or progress in intensity. When reassessment during a treatment session indicates that function is not normalizing or that symptoms are not decreasing, be alert to the need for further evaluation to determine a more appropriate technique, positioning, direction of force, or treatment intensity.

Elements of the OMT evaluation

OMT evaluation

A. **Screening exam:** An abbreviated exam to quickly identify the region where a problem is located and focus the detailed examination.

B. **Detailed exam:**

1. ***History***: Narrow diagnostic possibilities; develop early hypotheses to be confirmed by further exam; determine whether or not symptoms are musculoskeletal and treatable with OMT.
   - *Present episode*
   - *Past medical history*
   - *Related personal history*
   - *Family history*
   - *Review of systems*
2. ***Inspection***: Further focus the exam.
   - *Posture*
   - *Shape*
   - *Skin*
   - *Assistive devices*
   - *ADL*
3. ***Tests of function***: Differentiate articular from extraarticular problems; identify structures involved (see Chapter 3).
4. ***Palpation***
   - *Tissue characteristics*
   - *Structures*
5. ***Neurologic and vascular examination***

C. **Medical diagnostic studies:** *Diagnostic imaging, lab tests, electro-diagnostic tests, punctures*

D. **Diagnosis and trial treatment**

Through the physical examination the therapist correlates the patient's signs with their symptoms. A relationship between musculoskeletal signs and symptoms suggests a mechanical component to a problem that should respond well to treatment by manual therapy. The constellation of signs and symptoms revealed during the physical examination indicates the nature and stage of pathology and forms the basis of a treatment plan. For instance, before treating a patient who is unable to straighten a knee, you must first determine if the limitation is due to pain (e.g., lumbar radiculopathy), hypomobility (e.g., soft tissue contracture, intraarticular swelling, a meniscal block, nerve root adhesion), weakness (e.g., peripheral neuropathy, primary muscle disease), or a combination of these disorders.

■ Screening examination

The screening examination is an abbreviated exam intended to quickly identify the region of the body where a problem is located. It serves to define or focus additional examination and in some cases leads to a diagnosis and immediate treatment. The screening exam leads to one of the following three things:

» **A diagnosis** may be made if the physical signs are obvious, correlate well with the history and confirm your initial impressions;

» **Further detailed examination** may follow if insufficient data is collected and a diagnosis cannot be made;

» **Contraindications** to further examination or treatment may be uncovered and lead you to refer the patient to an appropriate specialist.

For experienced practitioners, there is no set sequence in which you perform screening examination procedures. The circumstances surrounding each particular problem determine how much and in which order you proceed.

Be careful not to over-examine, aggravate the patient's condition, or cause unnecessary pain during the screening examination. On the other hand, make sure you are thorough enough to gather all important information.

You must plan the examination from the very moment you meet the patient. And you must be prepared to modify your screening plan spontaneously based on emerging information during the process.

In practice, the screening exam is usually brief and results in either a provisional diagnosis or further, more detailed examination. It should give you a good idea of the type of problem and where it exists. If the diagnosis is still unclear or you wish to confirm your impressions, examine the patient further in the detailed examination (described later in this chapter).

Screening examination skills require mastery of the detailed examination in addition to much thought and clinical experience. Novice practitioners rarely conduct an efficient screening examination. We therefore recommend that novice practitioners first follow and master the detailed examination before relying heavily on screening examination findings.

Novice practitioners should first master the detailed examination before relying heavily on screening examination findings.

Components of the screening examination

Begin the screening exam by interviewing the patient for a brief **history** of the problem. You need enough information to determine where in the body to begin examination and which examination procedures will be most useful. A skillful examiner quickly gets the patient to describe their problem and the immediate circumstances preceding the onset. This brief history, if skillfully gathered and interpreted, can give you a description of the patient's symptoms and functional limitations, define the anatomical location of the problem, and identify any precautions.

Inspection begins from the very moment you meet the patient and start taking the history. Note static postures, respiration, and antalgic positions. The region to be examined should be visible so you can see swelling, discoloration, deformities, and skin changes. Observe the patient moving for valuable clues to the type and severity of their dysfunction. For example, watch how the patient gets up to move from the waiting room to the exam room and undress. These observations may lead you to ask further questions of the patient and guide you in planning further examinations.

The physical testing component of the screening examination varies, depending on the information obtained from the history. Use **active** and, if necessary, **passive movement** to further define the anatomical location and mechanical nature of the dysfunction. The emphasis in the screening exam is on the interpretation of active movements. Try to anticipate which movements will be painful so that they are not the first movements you test. If you provoke symptoms early or often in the examination you may make the rest of the exam difficult or impossible to interpret.

Use selected **resisted movements** to quickly screen muscle strength and the status of contractile elements and nerve supply. Since active, passive, and resisted movements can provoke symptoms, they give clues as to the structures at fault and the origin of symptoms. Perform additional **symptom localization** screening tests if you need to more clearly identify mechanical aspects of the problem.

The goal of **superficial palpation** in the screening exam is to quickly identify obvious changes in the characteristics of soft tissues or underlying joints. Palpation may confirm information obtained in the history or observed during inspection or active movements. Unsuspected information may also be uncovered which may require additional examination. For example, you might suspect neurological dysfunction if the patient does not feel your touch or is hypersensitive to palpation.

Superficial palpation sometimes leads to more specific examination using **passive joint and soft tissue movement** tests. Accessory joint mobility, stability, and pain are assessed with joint play movements. Passive soft tissue movements help assess the quality and texture of muscles, tendons, ligaments, and other soft tissues.

A **neurological or vascular exam** may be performed at any time during the screening exam, especially if some potentially serious condition is suspected which contraindicates further examination or treatment. For example, the patient may describe symptoms that suggest central nervous system pathology. In that case you might begin the screening examination with a neurological examination before any other test.

■ Detailed examination

A good patient history will often narrow diagnostic possibilities, however, an appropriate physical examination is still necessary to confirm the diagnosis.

Components of the detailed examination

1. History
2. Inspection
3. Tests of function (see *Chapter 3*)
4. Palpation
5. Neurological and vascular examination

History

During the history, you begin forming early hypotheses which subsequently must be confirmed or eliminated by further examination. In this way the history guides you in planning an appropriate physical examination.

a. Present episode
b. Past medical history
c. Related personal history
d. Family history
e. Review of systems

After obtaining the history, you should have in mind a list of possible diagnoses. Sometimes the history is so clear that you are confident of the diagnosis and, therefore, the physical exam may be brief and directed to confirm your impressions. On the other hand, the history may be so vague or confusing that many possible diagnoses must be explored. In this case the detailed exam must be broader in order to explore more possibilities.

If the physical examination does not confirm your initial impressions, proceed to further and more detailed questioning of the patient.

Present Episode

Obtain a complete description of all the patient's complaints and the events leading up to the current episode. Define any mechanical characteristics of the patient's complaint and identify cause-related or symptom-aggravating factors.

It is important to determine if the complaint is mechanical in nature. Non-mechanical symptom behavior raises the suspicion of more ominous diagnoses and may lead to a broader exam or referral to an appropriate medical specialist.

If the patient reports symptoms only during certain times, for example, in the evenings, it may be necessary to schedule the physical examination during that symptomatic period.

Symptoms (chief complaint):

- » Location: anatomical site or area of symptoms
- » Time: behavior of symptoms over a twenty-four-hour period
- » Character: quality and nature of symptoms
- » Influences: aggravating and alleviating factors
- » Association: related or coincidental signs and symptoms
- » Irritability: how easily symptoms are provoked and alleviated
- » Severity: degree of impairment and pain

History and course of complaint (chronology): Trace the chronology of relevant events leading up to the present episode.

- » Date of onset
- » Manner of onset: sudden, traumatic, or gradual
- » Pattern of recurrence: previous or usual manner of onset; related events; duration, frequency, and nature of episodes
- » Previous treatments and their effect

Past medical history

A complete medical history is especially important if you suspect the patient's problem is not musculoskeletal or mechanical in nature. Identify all major past health problems and recognize their possible relation to the patient's current complaints. Obtain the results of previous medical tests and treatments for further useful information. Remember that systemic and visceral diseases can mimic musculoskeletal disorders and their symptoms may even be temporarily alleviated with physical therapy procedures.

General health

» General health status

» Weight (recent weight loss or gain)

» Last physical examination (date and results)

» Medical tests (dates and results)

» Treatments, including medications (date, type, and effect)

Habits

» Sleep, diet, drugs (including coffee, alcohol, tobacco), activity level

» Major illnesses

» Hospitalizations, operations, injuries, accidents

Related personal history

Details about the patient's personal background and everyday environment may give insight into possible aggravating or complicating factors. Listening to a patient's typical day's or week's activities, especially occupational and recreational, often provides clear evidence as to the cause of the person's problem. Social, psychological, and financial hardships should also be considered, as they can greatly influence the success of treatment.

» Occupation (past and present work; future job requirements)

» Recreational activities

» Psychosocial status, including financial hardship

» Home environment (marital status, children ...)

» Typical day's activities

» Environmental factors (exposure to environmental pollutants ...)

Family history

Identify any patterns of recurring health problems in the patient's family or any possible genetic or familial conditions. Some joint and connective tissue disorders have a genetic cause or familial link.

» Age and/or cause of parents' and grandparents' deaths

» Hereditary, genetic and chronic diseases (parents, grandparents, siblings, children)

Review of systems

Answers to questions about each organ system and anatomical region can uncover symptoms not previously identified. A complete review of organ systems is especially important to rule out pathologies that might mimic musculoskeletal disorders in patients who come to the physical therapist without a medical referral. It can be difficult to determine whether symptoms are of visceral or musculoskeletal origin. For example, nerve root irritation in the thoracic spine can mimic symptoms of angina pectoris and make diagnosis difficult. The following systems should be reviewed:

» Integument (skin)

» Bones, joints, muscles

» Hematopoietic system

» Immune system

» Endocrine system

» Cardiovascular system

» Respiratory system

» Gastrointestinal system

» Genitourinary system

» Nervous system

» Lymph nodes

» Head

» Eyes

» Ears

» Nose

» Throat

» Mouth

» Neck

» Breasts

Inspection

Initial observations of the patient provide information which helps you further focus the exam. For example, watch the patient get up or down from sitting or take their shoes off for clues as to the body region where a problem exists. Make a mental note as to various areas of potential dysfunction and subsequently clarify these impressions with detailed examination.

Observe the patient both in static postures (static inspection) and while moving (dynamic inspection). The dynamic inspection includes selected daily activity movements and continue during other tests of function.

» Posture: habitual, antalgic, or compensatory body positions

» Shape: general body type, changes in normal contours, deformities, swelling, atrophy

» Skin: color changes, scars, callouses, trophic and circulatory changes

» Activities of daily living: gait, dressing, undressing, getting in and out of a chair

» Assistive devices: use of cane, crutches, corsets, prostheses

Tests of function

See *Chapter 3: Tests of function.*

Palpation

Palpation progresses from superficial tissues to deep structures and reveals asymmetries and deviations. Compare palpation findings in weight-bearing postures (standing, sitting) with findings in non-weight-bearing postures (lying). Some subtle palpation findings may only be detectable during activity. Palpation *during* many tests of function, *especially* passive movement testing, is therefore an essential part of an OMT evaluation.

Tissue characteristics

» Temperature
» Pulses
» Thickness
» Symmetry
» Crepitus
» Moisture
» Contour and shape
» Texture
» Tenderness
» Mobility and elasticity

Structures

» Nerves
» Fat
» Muscle
» Tendon
» Ligament
» Skin and subcutaneous tissue
» Tendon sheaths and bursae
» Fascia
» Blood vessels
» Bone

Palpation of the spine, pelvis, and ribs is difficult because the therapist must feel small articulations through deep layers of soft tissue, and asymmetries in the shape of bones and soft tissue are common. For this reason palpation findings indicating a positional fault are sometimes unreliable and should always be confirmed with specific mobility and localization tests.

Specific palpation of nerves follows the neural pathway, particularly at the most common sites of impingement.

Neurologic and vascular tests

Any suspicion of neurologic or vascular involvement should initially be considered a positive finding.

Neurologic tests

- » Deep tendon reflex testing
- » Strength and fatigability testing (including repeated resisted tests)
- » Sensory testing (light touch, pinprick, vibration, and position sense)
- » Tension signs and neural mobility tests
- » Girth measurements
- » (See *Volume II: The Spine* for additional neurologic tests relevant to spinal and central nervous system examination and treatment.)

No single neurologic test is sufficient to determine a diagnosis. Neurologic tests overlap other tests of function and must be interpreted in light of an entire constellation of signs and symptoms. For example, reduced strength can be a by-product of muscle, joint, or neurologic dysfunction. In addition, positive findings from any nerve test that involves limb movement (particularly root tension and mobility tests) can originate from a variety of tissues, including nerves, joints, and muscles. Separately assess the joints and muscles involved in each test and consider these when interpreting findings.

Vascular Tests

- » Pulses
- » Bruits
- » (See *Volume II: The Spine* for additional vascular tests relevant to spinal examination and treatment.)

Medical diagnostic studies

- » Diagnostic imaging (e.g., x-ray, CT scan, MRI)
- » Laboratory tests (e.g., analysis of blood and other body fluids)
- » Electrodiagnostic tests (e.g., EMG, EEG)
- » Punctures (e.g., biopsy, aspiration)

Diagnosis and trial treatment

A trial treatment is an essential evaluation tool. If examination findings implicate a joint condition which is treatable, confirm your diagnostic hypothesis with a trial treatment. If the patient's response to the trial treatment is as you predict, the diagnosis is confirmed. If the diagnosis involves shortened tissues, several trial treatments may be required before the diagnostic hypothesis can be confirmed.

Before initiating a treatment plan, you should be confident in your answers to the following questions.

- » Is there good correlation between the history and the physical exam?
- » What is the patient's diagnosis? What are their problems and priorities for treatment?
- » Do I have enough information to begin treatment or should I reexamine the patient?
- » Should I refer this patient for further evaluation?
- » Can I help this patient? What treatment do I have to offer?
- » Are there contraindications to treatment?

■ Notes

5 Joint mobilization

The mobilization techniques presented in this book evolved largely as a result of the following observations:

» One can *see* and *measure* decreased active movement of a limb and *feel* restricted joint play in the associated joint.

» Following treatment with passive translatoric movements, there is usually an increase in active movement, an increase in passive joint play, and decreased pain.

Joint mobilization is perhaps the most important component of OMT practice. Hands-on skill in joint mobilization enhances both diagnostic acumen and treatment effectiveness.

Joint mobilization

1. **Pain-relief mobilization**
   - *Grade I - IISZ in the (actual) joint resting position*
2. **Relaxation mobilization**
   - *Grade I - II in the joint (actual) resting position*
3. **Stretch mobilization**
   - *Grade III in the joint (actual) resting position*
   - *Grade III at the point of restriction*
4. **Manipulation**

While the subject of this book, and the emphasis in this chapter, is limited to joint mobilization, this is but one part of the larger scope of OMT practice. See *Chapter 6: OMT Treatment* for an overview.

Goals of joint mobilization

Mobilization treatment is based on a specific biomechanical assessment of joint hypomobility and hypermobility.

If the patient's symptoms are associated with an abnormal end-feel and a slight or significant **hypomobility** (Class 1 or 2), use Grade II relaxation-mobilization or Grade III stretch-mobilization techniques to improve joint function. Class 0 ankylosed joints are not mobilized.

If the patient's symptoms are associated with a slight or significant **hypermobility** (Class 4 or 5), apply stabilizing (limiting) treatment to normalize joint function. Complete instabilities (Class 6 dislocations or ligamentous laxity with instability) usually require surgical intervention.

Mobilization techniques

Pain-relief mobilization

Grade I - IISZ

If the patient has severe pain or other symptoms (e.g., spasm, paraesthesia) such that the biomechanical status of the joint cannot be confirmed or that Grade III stretching techniques cannot be tolerated, direct treatment toward symptom control. Symptom-control treatment should be applied only in the Slack Zone of the Grade I - II range.[1]

Grade I and II Slack Zone mobilizations, particularly intermittent traction movements, also help to normalize joint fluid viscosities and thus improve joint movement when movement is restricted by joint fluids rather than by shortened periarticular tissues.

Apply pain-relief mobilizations as *intermittent Slack Zone Grade I and II movements in the resting position or actual resting position.*

Pain-relief-traction mobilization

Grade I - IISZ

Intermittent Grade I and II traction-mobilizations in the Slack Zone, applied in the resting position or actual resting position (i.e., three-dimensional positioned traction), is the initial trial treatment of choice for symptom control.

Remember to apply mobilizations for pain relief within the Slack Zone, staying well short of the Transition Zone.

---

1 In some countries, practitioners refer to *Grade I and II Slack Zone mobilizations for pain relief* as "passive movements" and reserve the term "mobilization" for the treatment of hypomobility.

As soon as decreased symptoms allow the patient to tolerate full biomechanical testing with end-feel assessment, the focus of treatment can shift to the appropriate mobilization for hypomobility or stabilization for hypermobility.

Vibrations and oscillations

Short amplitude, oscillatory joint movements other than traction are also used for the treatment of pain. These movements are usually applied manually, but the use of mechanical devices such as vibrators may also be effective in the application of very high frequency and very short amplitude movement. These movements can decrease pain and muscle spasm, therefore improving mobility without stretching tissues.

Vibrations and oscillations can also be applied in the Grade IITZ and III range, interspersed with stretch mobilizations, to minimize discomfort.

■ Relaxation mobilization

Grade I - II

Relaxation mobilizations differ from pain-relief mobilizations in that they can be applied anywhere in the Grade I-II range, including both the Slack Zone and through the increasing resistance of the Transition Zone. It is important to differentiate relaxation mobilizations from the more gentle and benign Grade I-II pain-relief traction mobilizations which are applied only within the Slack Zone.

Apply relaxation joint mobilizations as *intermittent Grade I and II movements in the actual resting position* to decrease pain and relax muscles. Use them in cases where joint movement is limited by muscle spasm rather than by shortened tissues. Relaxation mobilizations are also useful as preparation for more intensive treatments (for example, a Grade III stretch mobilization) which can be more effective when the patient's muscles are fully relaxed.

Relaxation mobilizations should not produce or increase pain.

Following is a review of joint relaxation mobilization techniques. See Chapter 6 for a discussion of soft tissue relaxation mobilization techniques.

Relaxation-traction mobilization

Grade I - II

Apply intermittent traction-mobilizations in the actual joint resting position within the Grade I or II range, including the Transition Zone. Slowly distract the joint surfaces, then slowly release until the joint returns to the starting position. Rest the joint a few seconds in the starting position before you repeat the procedure. Between each traction movement readjust three-dimensional positioning (the actual resting position) of the involved joint as joint tissue response allows. You may need to interrupt the traction procedure and reposition the joint in different dimensions until the new actual resting position is found and repeated traction relieves symptoms. There should be a natural progression in joint position toward the resting position of the joint.

Avoid tissue stretching. Stay well within the Grade I and II range and do not mobilize into the Grade III range where tissue stretching occurs. Subtly and continuously modify joint positioning, mobilization forces, and the rhythm and amplitude of the traction procedure based on the patient's response to treatment.

Evaluate the effect of these carefully graded traction forces. You should observe an immediate improvement in signs and symptoms if your treatment approach is correct.

It is rare for Grade I or II intermittent traction to increase a patient's symptoms. If it does, you should:

» Adjust patient positioning. Continuously monitor changes in the actual resting position and adjust the patient's three-dimensional positioning as needed.

» Alter traction force. Early in the healing process a patient may tolerate only minimal forces.

» Correct an underlying positional fault. A positional fault can occur in both hypomobile and hypermobile joints. It is a condition in which joint partners are in an abnormal position, most often involving a hypermobile joint stuck in an unusual joint position. While minor positional faults often correct with a Grade II traction mobilization, strongly fixated positional faults may first need correction with a Grade III stretch glide-mobilization or manipulation.

» Discontinue traction treatment. In some cases, for instance with certain acute soft tissue lesions (e.g., ligamentous strain) traction treatment may be contraindicated along with any form of stretch to the injured fibers.

Stretch mobilization

Grade III

Grade III stretch mobilizations are one of the most effective means for restoring normal joint play. Stretching shortened connective tissues in muscles, joint capsules and ligaments can increase and maintain mobility and delay progressive stiffness and loss of range of movement in chronic musculoskeletal disorders.

Hypomobility presenting with a hard end-feel is characteristic of a bony limitation and should not be stretched. Restricted range of movement presenting with a normal end-feel is probably a normal anatomical variation, is rarely symptomatic, and is not stretched as a primary treatment. However, such "normal" joints may be stretched in order to release stress to a vulnerable neighboring hypermobile joint.

Grade III stretch mobilization is only indicated, and only effective, when a hypomobility is associated with an abnormal end-feel, is related to the patient's symptoms, and there are no contraindications.

Fixation of one joint partner is absolutely essential for an effective stretch mobilization.

Sustain a stretch mobilization for a minimum of seven seconds, up to a minute or longer, as long as the patient can comfortably tolerate the stretch. In viscoelastic structures, the longer a stretch is sustained the greater and more lasting the mobility gain. We instruct students to apply 30 to 40 seconds of stretch with the assistance of a mobilization belt in the larger joints. For greatest effect, continue the treatment for 10-15 minutes in a cyclic manner.

It is not necessary to release the joint completely between stretch mobilizations. A return to the end of the Grade II range, just easing off the stretch into the Transition Zone, is adequate before repeating the process.

Normally the time a stretch is sustained is more critical than the amount of force used. Poor gains in range are more commonly due to insufficient duration of stretch, rather than insufficient force. However, you must apply enough force to stretch the shortened tissue. To determine the most effective amount of force to use, begin with forces approaching, but not exceeding, what the patient safely tolerates during daily activities. In some larger joints, for example, in the shoulder, elbow, hip and knee joints, the force of Grade III stretch traction-mobilizations can be significant.

Grade III stretch mobilizations should not produce or increase the patient's dominant symptoms (chief complaint). However, a sensation of stretching in the form of slight local discomfort is a normal response to stretch-mobilization. A Grade III stretch mobilization should be discontinued if it produces protective muscle spasm, severe pain, or symptoms at locations other than the site being treated. Such a response to treatment suggests the need to reposition the patient, alter the intensity or direction of treatment, or to postpone stretch-mobilization until some healing occurs independently of treatment.

Grade III stretch-mobilizations usually produce immediate improvement within the first treatment session. You should see, hear, and feel a difference in the patient's dominant signs and symptoms. Lasting effects may require several treatments.

Preparation for stretch mobilization

Soft tissue dysfunction can alter joint movement and decrease the effectiveness of joint stretch-mobilizations. That is why treatment often begins with procedures to decrease pain and muscle spasm or increase soft tissue mobility. These adjunct procedures may also make the joint mobilization easier to perform and produce a longer lasting effect.

Treatment to improve circulation and thereby elevate soft tissue temperatures is useful preparation for Grade III stretch mobilizations. Warming tissues surrounding the joint prior to Grade III mobilizations makes them easier to stretch. Effective warming can be achieved by surface heat application or deep heat application (e.g., ultrasound, diathermy). However, the most effective way to "warm-up" tissues is with exercise.

The most effective way to improve circulation and "warm-up" soft tissues is with exercise.

It is reported that cooling tissues after stretch mobilization treatment helps preserve mobility gains for a longer period of time. We do not recommend cold application prior to or during stretch technique, since cooled tissues can be more easily injured from overstretching.

Progression of stretch-mobilization treatments

One of the most frequently asked questions, and also hardest to answer is, "How much treatment is enough?" The easiest answer is, "As much as necessary and as little as possible." Although the answer is clever and accurate it rarely satisfies students. I therefore provide the following general guidelines which are both conservative and safe. With experience, the nuances of clinical decision-making will become more apparent and you will find answers to these difficult questions.

If reassessment reveals increased range of movement or normalization of end-feel and decreased symptoms, then Grade III stretch-mobilization treatment may continue. If there is marked improvement in one treatment session, it is wise to discontinue additional treatments that day. Chronic cases and significant (Class 1) hypomobilities may require several treatment sessions before a change is apparent.

If reassessment indicates no change in mobility or symptoms, re-evaluate patient positioning and the vigor (i.e., time and force) and direction of treatment or reconsider whether mobilization is indicated at all, perhaps by referring the patient for further medical diagnostic evaluation.

Stretch-mobilization should be discontinued when gains in the patient's symptoms and range of movement plateau and the patient can perform active movement throughout this range.

It is important to stretch a joint in all restricted directions in which the joint would normally move. However, some stretch-mobilizations into some movement patterns and directions are safer, while other stretch-mobilizations have greater risk of patient injury and must be applied with skill and caution. In addition, a joint can be restricted in one direction (e.g., flexion) and hypermobile in another direction (e.g., extension). In this case mobilization may be indicated for the restricted flexion and contra-indicated for the hypermobile extension.

Novice practitioners should begin stretch mobilization treatments with a sustained traction-mobilization pre-positioned in the resting position (or actual resting position) and progressively re-position nearer and nearer to the point of restriction, as tissue response tolerates and allows. If the mobility gains produced by stretch-traction mobilization plateau, the practitioner may progress to stretch-glide mobilizations, first with the joint pre-positioned in the resting position, then progressing toward the point of restriction, just as for stretch-traction mobilization treatment.

Stretch mobilization is more effective and better controlled when joint stretching is carefully timed to occur during periods of maximum muscle relaxation. Reflex inhibition relaxation techniques such as PNF contract-relax and hold-relax techniques (i.e., active relaxation, post-isometric relaxation) and contraction of antagonists (i.e., reciprocal inhibition) can be very effective.

Stretch-traction mobilization

Grade III

A series of sustained Grade III stretch-traction mobilizations in the joint resting position is the recommended initial treatment for joint hypomobility. Apply stretch-traction mobilization at a right angle to the treatment plane.

Figure 5.1
Traction at a right angle to the treatment plane

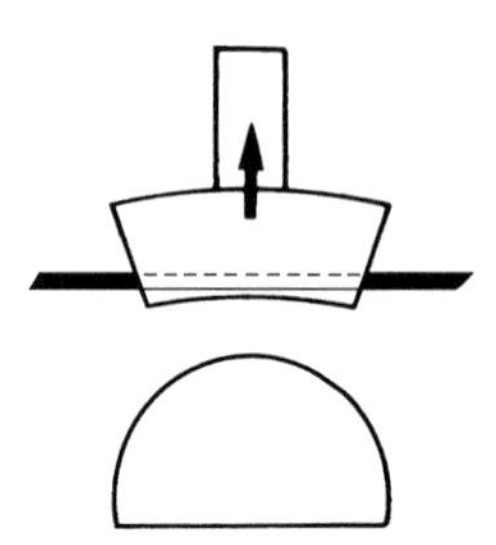

Grade III traction mobilization in the (actual) resting position can stretch any soft tissue that crosses the joint and limits joint movement, including muscle connective tissues, joint capsules and ligaments. As a trial treatment, apply about ten stretch-traction mobilizations. If reassessment reveals improvement, continue with this and progress toward the true resting position. Progress the stretch-traction mobilization in nonresting positions as improvement allows.

Grade III traction mobilization at the point of restriction is applied with the joint pre-positioned near the limit of range in the restricted movement direction. This maneuver will increase joint mobility primarily in the pre-positioned direction. For example, to increase a flexion restriction, pre-position the joint at the limit of the flexion range and apply the stretch-traction mobilization in that position. Skilled practitioners pre-position and stretch in more than one dimension, for example, in flexion with abduction (bi-axial joint) or flexion with abduction and external rotation (tri-axial joint). Progress the stretch-traction mobilization further into the restriction as improvement allows.

Treatment is often successful with skillful pre-positioning at the point of restriction combined with stretch-traction mobilization alone. However, in some cases, especially to treat the last degrees of restriction, it can be necessary to use stretch-glide mobilization as well.

Stretch-glide mobilization

Grade III

Stretch-glide mobilization directly stretches the tissues restricting joint movement. Progress to Grade III stretch-glide mobilizations if and when stretch-traction mobilizations no longer produce adequate mobility gains.

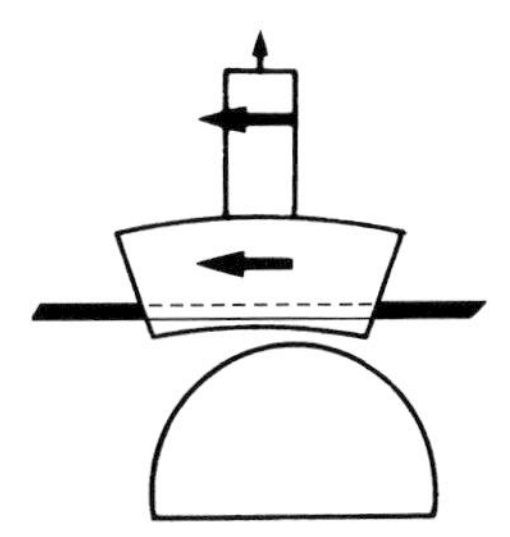

Figure 5.2
Gliding parallel to the treatment plane.

Progress joint pre-positioning in the same way as for stretch-traction mobilization. Start in the actual resting position, progress toward the true resting position, and gradually re-position the joint nearer and nearer to the point of restriction, as improvement allows. For best effect when the joint is pre-positioned at the movement limit, ease off the limit a little before applying the stretch-glide mobilization.

Apply stretch-glide mobilizations parallel to the treatment plane. Remember that when the *Concave Rule* applies, for example, with the fingers, elbow, toes, and knee, the treatment plane changes with each re-position of the distal (concave) joint partner. When the *Convex Rule* applies, the treatment plane does not change with each re-position of the distal (convex) joint partner, but remains with the stationary proximal concave joint partner.

Glide-mobilizations produce some intra-articular compression, more so with stiffer joints. To facilitate the glide mobilization and reduce these compressive forces acting on the joint, combine it with a Grade I traction movement. In joints with advanced degenerative changes, or which are painful when compressed, it may be necessary to use additional traction force in order to apply the glide-mobilization without pain.

The most effective mobilizations stretch a joint in the direction of most restricted gliding. However, if performed with poor technique or with excessive force they can injure sensitive joint structures. Stretch-glide mobilization in a severely restricted gliding direction (Class 1 hypomobility) may produce joint compression and be too painful for a patient to tolerate. In this case, return to Grade III stretch-traction mobilizations carefully applied in less restricted and less symptomatic positions. Once mobility status improves to a slight hypomobility (Class 2), progress again to specific Grade III stretch-glide mobilization in the most restricted gliding direction.

Rotation mobilization

We do not teach rotation joint mobilizations around a longitudinal axis because they can produce significant compressive forces with adverse effects.

The safest way to increase joint rotation range is to use a Grade III stretch-traction mobilization in conjunction with specific three-dimensional positioning. Pre-position the specific joint at the point of its restricted rotation, and then apply a Grade III traction mobilization at a right angle to the joint treatment plane. If this procedure does not completely restore the rotation movement, progress to a linear stretch-glide mobilization at the end-range of the restricted rotation, with a simultaneous Grade I traction force to protect the joint.

The skilled application of three-dimensional stretch-traction mobilization or specific stretch-glide mobilizations is safer and, in skilled hands, just as effective as rotation mobilization.

■ Manipulation

We practice manipulation as a high velocity, small amplitude, linear movement in the actual resting position, applied with a quick impulse ("thrust") to a joint showing a suitable end-feel, to effect joint separation and restore translatoric glide.

You must understand the indications for, and especially the contraindications to, manipulation in order to prevent patient injury. OMT training follows the guidelines of the International Federation of Orthopedic Manipulative Therapists (IFOMT). IFOMT guidelines recommend a specific sequence of education that begins with extremity joint mobilization and progresses to extremity joint manipulation, before the practitioner begins to learn spinal manipulation. Training in extremity joint manipulation, particularly the relatively safe traction manipulations applied in the joint resting position, can begin early in OMT education. However, spinal manipulation is advanced and should be used only by those with long-term training and clinical supervision.

While effective in skilled hands, manipulation also carries risk of serious injury.

Risks to the patient increase with rotatory manipulation techniques, especially in the cranio-vertebral region. In an attempt to reduce the risks inherent in manipulation, we have worked for many years to perfect techniques which use a translatoric linear traction-thrust, rather than a rotatory-thrust. While this type of linear thrust is technically more difficult to perform, it is far safer and equally effective. IFOMT recommends manipulative techniques which "eliminate rotary stresses and emphasize glide and distraction movements." We no longer teach rotatory manipulation techniques either for the extremity joints (since 1979) or for the spine (since 1991).

■ Joint compression in treatment

I do not teach joint compression techniques because they can too easily aggravate a joint condition. However, some practitioners believe that passive manual joint compression can stimulate cartilage nutrition and regeneration and use it for that purpose, particularly in certain extremity joints.

Little is known about the physiological effects of manual joint compression treatment or whether an interspersed traction component is essential for its efficacy. Critical to the maintenance of articular cartilage is its fluid supply of nutrients by diffusion. This fluid nutrient transfer is facilitated by *changes in joint loading* which create pressure changes. Therefore, it has been hypothesized that compression may be a useful joint mobilization technique. Following the same logic, our intermittent traction approach may also provide the necessary pressure changes, thus facilitating articular cartilage nutrition.

Rolling, gliding, and compression are physiological stresses joints experience with normal movement. In fact, these stresses are *necessary* for the maintenance of articular cartilage. When there is an imbalance of rolling, gliding and compression, joints begin to show the effects of wear and tear, marking the onset of degenerative joint disease (DJD). For example, too much compression may occur with excessive running or jumping activities which can lead to DJD. On the other hand, not enough stress to the joint, as with prolonged immobilization in a cast or bed rest, can also lead to degenerative joint disease.

If joint compression occurs during a patient's treatment program, the amount of load-bearing is increased gradually and monitored closely to avoid pain. Therapists use standard protocols for graduated return to full weight-bearing in the lower extremity joints. The progression usually begins with toe-touch weight-bearing using two crutches and progresses to one crutch, then a cane, and eventually full weight-bearing. Another common progression starts with active assisted movement, then active movement, and finally resisted movement. These progressions represent a kind of *graduated compression therapy* which the patient controls based on their tolerance to the activity. Premature load-bearing treatment can lead to joint swelling and additional injury to the patient.

Many seemingly benign daily activities produce joint compression and can aggravate a patients symptoms. For example, sidelying induces significant compression both through the shoulder girdle joint complex, in the cervical and upper thoracic spine, and in the hip joint. Management of this patient would likely include instruction in how to position pillows under the thorax and neck to reduce shoulder and spinal compression during sidelying sleep. Management of this patient would also likely *avoid* additional joint compression during manual treatment.

We are aware that our gliding techniques often also have a compression effect, especially at the end range of motion. Traction and glide mobilization techniques are safer than joint compression techniques, and may very well provide the needed pressure changes to facilitate articular maintenance.

6 OMT treatment

Elements of OMT

Mobilization is but one part of OMT treatment and is often more effective when supplemented with other procedures and modalities. The sequencing of these adjunctive procedures can greatly influence the outcome of mobilization techniques. For example, a stretch mobilization preceded by heat application often produces greater mobility gains, and ice application and specific home exercise following a stretch mobilization can better preserve mobility gains.

While you study the following treatment guidelines, remember that clinical experience, not theory, is the most important criteria upon which treatment decisions are made.

OMT Treatment

A. To relieve symptoms
   1. *Immobilization*
   2. *Thermo-Hydro-Electro (T-H-E) therapy*
   3. *Pain-relief mobilization (Grade I - IISZ)* (see Chapter 5)
   4. *Special procedures*

B. To increase mobility
   1. *Soft tissue mobilization*
      a. Passive soft tissue mobilization
      b. Active soft tissue mobilization
   2. *Joint mobilization* (see Chapter 5)
      a. Relaxation mobilization (Grade I - II)
      b. Stretch mobilization (Grade III)
      c. Manipulation
   3. *Neural tissue mobilization*
   4. *Specialized exercise*

C. To limit movement
   1. *Supportive devices*
   2. *Specialized exercises*
   3. *Increasing movement in adjacent joints*

D. To inform, instruct, and train

Treating related areas of impairment

In addition to treating the primary joint lesion, the manual therapist also evaluates and treats related areas of impaired function. For example, knee lesions can be associated with dysfunction in the tibio-fibular joint or hip; shoulder joint lesions can be associated with dysfunction in the acromio-clavicular joint and mid or lower cervical spine; hip joint lesions can be associated with dysfunction in the pelvis or lumbar spine. Remember also that peripheral joint pain can be of spinal origin (refer to the dermatome, myotome and sclerotome charts in *Volume II: The Spine*).

Reassessment

Reassessment is important at the beginning and end of each treatment session as well as during the treatment session. If retesting reveals increased range of motion or decreased pain, then treatment may continue as before. If retesting reveals a *marked* improvement in range of motion, I advise novice practitioners to stop treatment for that day and continue the treatment on a subsequent day. I make this recommendation because novice practitioners all too often overtreat the patient in the mistaken belief that "more is better."

Under no circumstances should treatment result in discomfort or pain which persists beyond the day.

■ Treatment to relieve symptoms

Symptom control treatments can be indicated for both hypermobile and hypomobile joint conditions and in the presence of nerve root findings. Use symptom control techniques when:

» severe pain or other symptoms (for example, an empty end-feel) interfere with biomechanical assessment of the joint

» end-range-of-movement treatment is contraindicated or cannot be tolerated (e.g., in certain stages of disc pathology)

» inflammatory processes, disc pathology, or increased muscle reactivity around a symptomatic joint decrease gliding movement and restrict functional movement without structural soft tissue shortening (e.g., in the presence of normal muscle length or a normal or even a lax joint capsule)

In cases where nerve root irritation or the status of the intervertebral disc interferes with assessment of the biomechanical status of the joint (for example, due to severe pain or spasm), or when the nature of the condition does not allow for biomechanically based treatment, direct treatment toward symptom relief.

Immobilization

With some clinical conditions, immobilization is appropriate and necessary for a prescribed time. Selecting the correct general or specific immobilization method as well as *timing when and how long* to immobilize is important to the success of treatment. Acutely severe, painful and inflammatory conditions, instabilities, and recent post-surgeries may benefit from a prescribed duration of immobilization. General bed rest may be the only alternative with certain painful, inflammatory conditions, especially in the weight-bearing joints. Specific immobilization methods such as the use of casts, splints, braces, and taping can be used to protect a joint while the patient continues to function.

Thermo-Hydro-Electro (T-H-E) therapy

The judicious use of various forms of cold, heat, water, or electrotherapy can be an effective means to modulate pain, enhance relaxation, and reduce swelling. Integrated with manual therapy, modalities are used in preparation for mobilization and afterwards to prevent or limit treatment-related soreness. As with all treatments, selecting the correct technique, and determining when and how long to use it, is critical.

Pain-relief mobilization (Grade I-II SZ)

See *"Pain relief mobilization"* in *Chapter 5: Joint Mobilization.*

Special procedures to relieve pain

Acupuncture, acupressure, and various forms of soft tissue mobilization have long been used for pain relief through reflex pain modification, inhibition of muscle spasm, and the reduction of swelling. These are safe treatments even in the presence of serious musculoskeletal dysfunction.

■ Treatment to increase mobility

Soft tissue mobilization can facilitate Grade III stretch mobilization by loosening tight soft tissues that limit joint movement. In practice, treatment often begins with soft tissue treatments such as functional/pumping massage and muscle stretching to increase soft tissue mobility. In some cases, particularly with chronic disorders, both periarticular tissues and muscles are restricted near the same point in the range. In such cases it is necessary to alternate Grade III stretch joint mobilization with soft tissue mobilization or muscle stretching and to take care not to move joints beyond their natural or actual range of movement during the soft tissue procedures.

Soft tissue mobilization

Whether or not a particular technique is viewed as soft tissue mobilization depends on the viewpoint of the clinician. Soft tissue treatments can affect many structures including joints, nerves and blood vessels. What distinguishes the soft tissue treatment from other forms of treatment is that the clinician uses soft tissue assessment to monitor change. The intention is to change soft tissues. Assessment is made by monitoring soft tissues. The clinician continuously monitors tissue response and instantaneously modifies treatment.

Good manual soft tissue technique requires sensitivity to constantly fluctuating patient responses. The clinician must recognize these subtle changes and immediately and continuously modify the treatment.

Just as joint movements are classified as either translations (i.e., joint play accessory movements) or rotations (i.e. physiological bone movements), so are soft tissue movements.

Accessory soft tissue movements ("muscle play") cannot be performed actively. Friction massage, a passive lateral movement of muscle, is one example of muscle play.

Physiological soft tissue movements can be performed actively or passively. Traditional muscle stretching, and the lengthening and shortening movements that occur with muscle contraction and relaxation, are examples of physiological soft tissue movements. Treatment using physiological soft tissue movements generally utilize limb movement (bone rotations) to alter soft tissue tension.

Some forms of soft tissue mobilization such as functional/pumping massage are most effective when we allow the underlying joints to move as well. We often encourage and guide underlying joint movement by using a coupled movement pattern during soft tissue mobilization.

Soft tissue mobilization techniques can be broadly classified according to the amount of patient participation as either passive or active. The level and type of patient participation to use is an important clinical decision. Patient participation can vary from none at all, to the patient supplying most of the mobilizing force. Patient participation depends on many factors, including the chronicity and painfulness of the problem as well as the patient's willingness and ability to move.

Passive soft tissue mobilization

During passive soft tissue mobilization (STM) the patient does nothing but relax while you provide all the movement and force. This method is especially useful for soft tissue approximation or shortening. These are appropriate for treatment of certain acute soft tissue injuries where the objective is early movement with minimal tissue elongation or stretching. However, this approach may not be effective if the patient has difficulty relaxing while they are passively moved. There are many forms of passive STM, including classical massage, functional massage (Evjenth), and friction massage (Cyriax).

Active soft tissue mobilization

Contract-relax followed by passive physiological lengthening of soft tissues (muscle stretching).

Following a muscle contraction there is a brief period of relaxation when the muscle can be more easily stretched. During the relaxation phase, the practitioner stretches the soft tissues by moving muscle attachments maximally apart and holding them there. This kind of passive stretching can be uncomfortable and even painful in the stretched tissues, but should not increase the patients primary symptoms. The patient must be able to relax despite discomfort. Refer to the books by Evjenth and Hamberg for the definitive description of these muscle stretching techniques.[1]

Contract-relax followed by passive accessory mobilization of soft tissues.

Following a muscle contraction there is a brief period of relaxation when the muscle can be more easily mobilized. During the relaxation phase, the muscle can be passively moved in a variety of ways depending on how the muscle responds. The practitioner times the soft tissue mobilization to take full advantage of the relaxation period. This technique is useful for passive manipulation of a muscle in cases where the muscle will not easily relax.

Contract with simultaneous mobilization of soft tissues.

The practitioner uses resistance to guide the patient's movement in order to actively elongate specific muscles. Simultaneously, the practitioner passively manipulates the antagonistic muscle. An example is manipulation of the hamstring muscles while simultaneously resisting knee extension (quadriceps activation). This technique takes advantage of the neurological phenomena called "reciprocal inhibition" and can be quite strong. This is useful when patients have difficulty relaxing while they are passively moved. It is also useful for more forceful or vigorous stretching. Patients seem to tolerate this technique well, perhaps because they control much of the force.

---

1 See Evjenth and Hamberg, *Muscle Stretching in Manual Therapy, Volumes I and II,* 1984 Alfta Rehab Forlag, Sweden, for a description of muscle stretching techniques.

Passive stretching principles

Integrate passive stretching with active soft tissue relaxation techniques whenever possible. Before stretching, test muscle length, nerve mobility, end-feel, and the underlying joints to make sure stretching is indicated and safe.

» To test muscle length, position muscle attachments maximally apart, taking into consideration both primary and secondary muscle functions.

» Determine that shortened muscles, and not a joint stop, is limiting movement.

» Examine underlying joints to insure they can withstand the stresses imposed on them during stretching. Stretching muscles over joints that are unstable, inflamed, or have decreased joint play can result in their injury.

When stretching muscles, observe the following principles:

» Warm the muscle prior to stretching, with exercise or passive heat applications, to facilitate relaxation.

» Precede stretching with an isometric contraction of the muscle to be stretched to obtain maximal relaxation.

» If the muscle crosses more than one joint, apply the stretch movement through the least painful, most stable, and largest joint.

» It is generally more effective, and comfortable for the patient, to stretch using a lower force sustained for a longer time (60 seconds or more) than greater force for shorter time. Applying stretching force for a longer time is more likely to result in plastic deformation of soft tissues rather than the more temporary elastic changes.

Joint mobilization to increase mobility

See *"Stretch mobilization"* in *Chapter 5: Joint mobilization.*

Neural tissue mobilization

In cases where an overt or suspected nerve root condition is accompanied by severe symptoms, treatment often begins before the physical evaluation is complete. The neurological examination should still be performed, possibly with creative application of each test maneuver in the patient's symptomatic postures. For example, if the patient reports symptoms when standing, and not when lying down, then the examination procedure may only test positive when the patient stands. Defer less critical biomechanical joint assessments and physical examination maneuvers that could risk further injury until the patient can tolerate them safely.

Intermittent traction is the safest and often the most effective treatment for nerve root lesions. **Grade I and II traction mobilization** can reduce nerve root irritation by improving metabolic exchange via the vascular system and by improving drainage of waste products from the inflamed nerve tissue. Apply a trial treatment with intermittent traction as for the patient with severe symptoms, first within the Grade I and II range, but with more frequent reassessment of neurological status (e.g., key muscle strength and reflexes, tension signs, nerve mobility) during and between traction maneuvers. Continuously monitor changes in the patient's actual resting position and adjust three-dimensional joint positioning as changes take place in the involved joint. Other symptom control procedures may also be useful.

In cases where nerve root symptoms are associated with segmental hypomobility, progress the traction to a **stretch-traction mobilization (Grade III)** with three-dimensional positioning. Grade III stretch traction mobilization can improve the spacial relationships between the involved structures, adapt the nerve root to a new tension relationship, and in some cases, improve disc and neurostructural placement.

Once nerve root findings are no longer dominant, progress treatment to other procedures for any associated hypomobility or hypermobility. Because spinal rotation-mobilizations (around the longitudinal axis) can aggravate a nerve root condition, avoid them in patients with a history or suspicion of nerve root involvement.

In certain clinical situations when joint and soft tissue mobilization techniques have not succeeded in alleviating symptoms, neural tissue mobilization may be indicated. There are specific techniques for mobilizing nerves in relation to their perineural tissue which, when appropriately applied, can be effective. *I do not recommend these techniques for the novice practitioner as they may involve the provocation of neurological symptoms,* and I do not discuss neural tissue mobilization techniques in this book.

Specialized exercise to increase mobility

The therapeutic application of exercises is the cornerstone of physical therapy. Almost all physical therapy patients should have exercise as part of their treatment program. Exercise should begin as early as possible and each patient should have a home exercise program.

No uniform regimen of exercise is applicable to all patients with hypomobility. Just like mobilization, exercise should be specifically tailored for the individual. We do not recommend the routine issue of preprinted exercise protocols based on medical diagnosis rather than examination findings. For exercise to effectively complement mobilization, it must be administered by the same clinician providing the mobilization treatment and not delegated to some other practitioner as an afterthought.

Automobilization (self-mobilization) exercise is useful for all patients with joint hypomobility to maintain or increase mobility. Automobilization exercises should be tailored to each individual's needs. For example, while some patients with restricted lumbar lordosis may benefit from spinal extension exercise, there are many patients whose symptoms worsen with spinal extension exercises, including those with spondylolisthesis, kissing spines, stenosis of the spinal canal, or with pain from working in prolonged extension postures.

In patients with both hypomobility and hypermobility in nearby spinal segments, the patient may need stabilization training to protect the hypermobile area during mobilization exercise for the hypomobile area. (See also *Autostretching* by Olaf Evjenth and Jern Hamberg.)

■ Treatment to limit movement

Hypermobile joints are often misdiagnosed as hypomobile and therefore mismanaged by practitioners unskilled in passive movement testing. Misdiagnosis is common when hypermobile vertebrae, especially a significant hypermobility (Class 5), gets "stuck" outside of its normal resting position (i.e., in a positional fault). The skilled application of traction and gliding test maneuvers sometimes releases the joint and clearly reveals the underlying hypermobility. In other cases, the positional fault may need correction with Grade III stretch-glide mobilization or manipulation before the underlying hypermobility becomes apparent. The nature of the end-feel determines whether the hypermobility is a normal anatomical variation (and should not be treated) or whether it is pathological (and might benefit from treatment).

The management of hypermobility limits or minimizes joint movement in the excessively mobile directions. This is accomplished in four ways, often concurrently, by: 1) specialized exercises, 2) increasing movement in kinetically related (i.e., adjacent) stiff joints, 3) taping, orthoses, and other supportive and controlling applications, and 4) instruction in body mechanics and ergonomics. Hypermobility treatment is a long-term process and requires persistence and patience from both patient and therapist.

Grade III stretch mobilization is contraindicated for hypermobile joints.

Supportive devices

Supportive devices such as lumbosacral belts and cervical collars can help to protect involved joints during an acute stage. These devices can also be used after treatment is completed when the patient works in unusual postures, during prolonged activities such as sitting, while playing sports, or if symptoms are recurrent. Most often lumbar belts are made of elastic material to minimize the muscle wasting associated with prolonged rigid immobilization. They are only used if needed and are always supplemented with strengthening exercises.

In more serious and chronic cases, a rigid support may be necessary (e.g., body jacket, leather corset). In these cases, a strengthening program (usually isometric) is essential to counteract the deconditioning that accompanies rigid immobilization.

Specialized exercises for hypermobility

Specialized muscle training is necessary to limit and control excessive movements. It is common for the small one- and two-joint spinal muscles (i.e., multifidus, rotatores) to be atrophied from disuse at a hypermobile segment. Controlled contractions of these muscles, first facilitated by the manual therapist and later continued with autostabilization exercises by the patient, can be an important first treatment step.

Patients with hypermobility must also change any habitual motor behaviors that stretch a vertebral segment in a hypermobile direction. This usually involves a long-term movement reeducation program emphasizing coordination and kinesthetic retraining in a variety of functional postures (including lying, sitting, standing) until the patient can demonstrate safe behaviors in timing, recruitment, and intensity of muscle activity around the hypermobile segment.

Slight hypermobilities (Class 4), while often asymptomatic, are still at risk for overstretching injuries during activities that place the joint at end ranges of movement and can progress to a symptomatic (Class 5) hypermobility. For this reason, specialized muscle training and ergonomic instruction are important whether or not the hypermobility is symptomatic.

Increasing movement in adjacent joints

Increasing movement in adjacent joints will decrease movement forces through the hypermobile joint during functional activities and will increase the opportunity for a hypermobile segment to heal and stabilize. For example, a hypermobile lumbar segment will be stretched less often and less forcefully during daily activities if the adjacent thoracic and lumbar spinal segments and the hip joints can contribute their full range of movement to a given activity. Movement in joints proximal and distal to the hypermobile segment can be enhanced with joint and soft tissue mobilization, automobilization, and other specialized exercises. Mobilize adjacent hypomobile joints as soon as possible, even if they are asymptomatic.

■ To inform, instruct, and train

Patient education takes time, but often saves time in the end as it leads to active participation by the patient and clearer communication between patient and health care provider. Many disturbances of the locomotor system are chronic, recurrent conditions which require self-management by the patient both at home and at work. Our manual therapy system stresses the role of the patient in reestablishing and maintaining normal mobility, in preventing recurrence, and in improving musculoskeletal health.

In addition to home exercises, we instruct patients in activities of daily living (ADL), body mechanics, and ergonomics. Instruction should be given not only in home exercise, but in methods for pain relief, for example traction, ice, heat or taping.

Home instruction is especially important if the patient's activities exacerbate neurological symptoms. Patients can be taught how to monitor their neurological signs and use them as a guide to determine safe activity levels.

Patients need instruction in what postures and movements to avoid and in developing new and more healthful ways of moving and working. Training programs emphasize coordination, kinesthetic retraining, strength, and endurance until the patient can demonstrate consistent and safe behaviors in timing, recruitment, and intensity of muscle activity during a variety of functional activities.

Therapeutic training can be provided on an individual basis, or in groups (e.g., back school). Ideally, patients will continue their training even after discharge from formal treatment, preferably at a facility with physical therapists as training instructors.

■ Research

Many challenges confound the conduct of useful research in the manual therapies. The validity of clinical trials is complicated by the many variables which confound accurate determinations of cause and effect in musculoskeletal disorders, and by the difficulties in developing valid measurement tools for manual interventions. Work is ongoing in the areas of inter- and intrarater reliability studies for manual techniques, however, all too often a manual therapy novice performs the manual techniques in a research study, rather than a master practitioner. This will, of course, impact the research results. There is also much work to be done in the development of accurate and meaningful functional diagnoses and assessment measures for monitoring changes in patient response. For researchers with a pioneering spirit, creativity, and determination, this is indeed an exciting new arena for study.

TECHNIQUE

7 Technique

Learning manual techniques

It takes years of study and practice to achieve mastery in Orthopedic Manual Therapy. Just as with mastery of a musical instrument, the theory and basic technique can be learned quickly, but it takes years of practice to play well.

Practitioners new to manual therapy are often dangerously heavy-handed. It may take much practice before a practitioner can reliably sense when they are approaching the first stop and can accurately sense the end-feel. To attempt a Grade III stretch-mobilization before mastering this skill runs the risk of injuring the patient or student practice partner with overstretching or unwanted compression forces.

Novice practitioners should first master soft tissue techniques and joint testing techniques, especially Grade I and II movements, before attempting Grade III stretch-mobilization techniques. When practicing mobilization on asymptomatic subjects, we recommend students use only within-the-slack Grade II mobilization forces to avoid tissue injury or joint overstretching.

One cannot learn orthopedic manual therapy from books and classroom teaching alone. Students must take the time to observe the intricacies and effectiveness of treatment delivered by a master clinician and must work to develop their own manual skills in a supervised clinical setting with real patients.

Learning specific manual mobility testing

Joint movement tests are an excellent method for monitoring change in a patient's physical status and for assessing a patient's response to treatment. But the technique is only as good as the therapist using it. The skill to feel and judge specific joint movements takes time, talent, and frequent practice.

We find that the practice of soft tissue treatments, especially functional massage, helps develop passive movement skills. After some time working with soft tissues, you will begin to feel the presence of bones and joints beneath the soft tissues and how these structures move. Later you will develop the ability to judge how much these structures move in relation to each other and whether the quality of movement is normal.

Applying manual techniques

A written description of a manual technique cannot adequately address the many nuances in patient handling that are critical to effective practice. For this, supervised clinical practice is essential. However, certain principles are prerequisite to the skilled application of manual techniques. Application of these principles will ensure efficient and safe use of the therapists's body and effective treatment for the patient.

Variations in functional joint anatomy

Generally, if joint play end-feel is normal, the joint is normal, regardless of asymmetries or deviations from established norms in range or direction of movement.

There is considerable normal anatomical variation from individual to individual, and considerable asymmetry from one side of the body to the other within an individual. The skilled OMT practitioner makes treatment decisions primarily on the basis of abnormal quality of movement, not on printed norms for movement.

For example, during joint play testing of the acromioclavicular joint you may discover that the concave joint surface of the acromion faces more medially on one side of the body and faces more laterally on the other side of the body. Or you may discover that, while your patient's total range of internal rotation and external rotation is equal for both shoulders, there is 20° more external rotation on the right and 20° more internal rotation on the left with all *normal end-feels*. Such findings are likely the result of asymmetrical orientations of the glenoid fossas, rather than joint pathology. Years of participation in an activity which is asymmetrical can also lead to asymmetrical adaptations in anatomical structure, for example sports such as tennis, golf and javelin.

If joint play end-feel is normal in all directions, the joint is normal, regardless of asymmetries or deviations from established norms in range or direction of movement.

Objective

The difference between a joint testing technique and a joint treatment technique is not always obvious. Joint play testing techniques can also be applied in the resting position as gentle Grade I and II traction mobilizations for pain relief or relaxation. Grade III stretch-mobilization techniques can sometimes also be used for symptom localization and end-feel testing.

With changes in grip, fixation, and positioning, many joint mobilizations can be adapted for use as a test, as a treatment for pain relief and relaxation, or as a stretch-mobilization. In addition, with changes in joint position the effect of the test or treatment can be much more specific. In the following chapters, we suggest the best application for each technique in its title:

» **"Test"** indicates that the technique is usually used for testing only. We illustrate linear, translatoric tests with straight arrows. We also indicate whether the objective of the test is for "mobility and symptom screening" or to "evaluate segmental range and quality of movement, including end-feel."

» **"Test and Mobilization"** indicates that the technique can be used for testing joint play (Grade II), for testing end-feel (Grade III), and also for stretch-mobilizations (Grade III). Both test and mobilization procedures usually use the *same grip*. "Test and Mobilization" *traction* techniques in the resting position can also be applied for pain relief (Grade I and IISZ) or muscle relaxation (Grade I through IITZ).

» **"Mobilization"** indicates that the technique is adapted with *alternate grips* or stronger fixation (for example, with straps) for more effective stretch-mobilizations (Grade III).

The technique objectives outlined in this basic book are guidelines only. Skilled practitioners will adapt and modify the techniques as the patient's condition and treatment goals dictate.

■ Starting position

Patient's position

Techniques should be applied in a sequence that is efficient and requires a minimum of patient repositioning.

First, place the patient's body in a position of comfort to encourage relaxation and minimize muscle tension, then position the specific joint(s) to be mobilized.

For most evaluation and basic mobilization techniques, position the patient so that the involved joints are in the resting position or in the actual resting position. In these positions the muscles surrounding the involved joint usually also relax. However, repeated trials may be necessary to find the best starting position, for example, the actual resting position for pre-positioned pain-relieving, three-dimensional traction.

» **If the patient is in a sitting position** the feet should be supported on the floor to contribute to the stability of the body necessary for proper positioning of the spine during evaluation and treatment.

» **If the patient is prone** it is usually necessary to place an appropriately sized pillow under the patient's stomach (even if the patient has a protruding abdomen) to position the lumbar spine in a comfortable position. A pillow may also be necessary under the thorax to maintain a resting position there. In some cases it is necessary to lower the head piece of the treatment table in order to achieve adequate muscle relaxation.

  The head piece of a manual therapy treatment table should have an opening for the patient's nose and mouth so they need not rotate their necks in order to breath. Cervical rotation increases tension of the cervical and shoulder girdle muscles.

» **If the patient is sidelying** the hip and knee joints should be flexed to provide stability. In sidelying, the patient's position should approximate the normal spinal curvatures observed in standing. In many cases, especially with females with a broad pelvis, it is necessary to place a pillow or a roll under the patient's waist for comfort.

» **If the patient is supine** the patient's head should be supported directly by the table or by a pillow, and the patient's legs should be slightly abducted and relaxed. For comfort and relaxation, it may be necessary to place a pillow under the patient's knees, to have the patient in a hooklying position, or to place a positioning pillow under the lumbar area.

The therapist must often modify some other positions to accommodate the characteristics and flexibility of individual patients.

Therapist's position

It is important that you assume an ergonomically and biomechanically sound posture as close as practical to the patient. Such a posture requires a wide base of support, flexed hips and knees, and natural lumbar lordosis. Adjust the height of the treatment table to ensure efficient and effective body mechanics.

■ Hand placement and fixation/stabilization

During most basic joint test and mobilization techniques, you move one hand with the patient's distal joint partner and keep the other hand stable for palpation, stabilization or fixation. Both your moving hand and your palpating/stabilizing hand monitor the quality and quantity of movement.

Grip

Grips for testing maneuvers and gentle Grade I and II mobilizations differ from grips for longer duration stretch-mobilizations. Grips for testing and gentle mid-range mobilizations use a smaller contact surface, sometimes using only your fingers for the grip. Grips for longer duration stretch-mobilizations use the broader contact surfaces of your hand along with more efficient therapist body mechanics and stronger fixation. In larger joints the grip may be reinforced with straps or with your body.

The less contact pressure the manual therapist uses, the more sensitive the therapist's hands are for monitoring movement quality. Since only a small degree of linear movement is available in any individual joint, excessive contact pressure can reduce movement, mask feedback about movement quality, distort the movement, and even elicit muscle guarding.

In practice, a well-placed grip close to the joint space of two adjacent joint partners, can also produce a Grade I traction sufficient to neutralize, or decompress, the joint and thus facilitate the test or mobilization procedure.

Modify and adjust your grip for patient comfort. For example, it may be necessary to push aside sensitive soft tissue structures such as nerves, muscles, or tendons. Or you may need to adjust your grip away from tender bony prominences.

The skilled manual therapist should be able to perform stabilizing/fixating and moving/mobilizing functions equally well with either hand, from either side of the patient. The techniques in this book are accompanied by photographs (figures) that show a technique after it has been performed, i.e., in the terminal position. To perform the same technique on the opposite side of that shown in the picture, simply stand on the opposite side of the patient and switch your stabilizing and moving hands. Students should practice testing and mobilization techniques on both sides to train both hands for both functions.

Therapist's stable hand

With many mobilization techniques, the practitioner keeps one hand stable while the other moves. Your stable hand provides fixation and is usually positioned just proximal to the joint space. The fingers of your stable hand are also used to palpate the joint space. It is much easier to palpate movement in a joint if your palpating finger is stable and not moving.

During most specific passive joint function tests and some mobilizations, the practitioner palpates with one finger of the stable hand. Most therapists use the index finger as the palpating finger (as illustrated in the photographs in this text), but individual therapists may find another finger more sensitive or more comfortable to use. Position your palpating finger at the targeted joint space with contact to both joint partners. (In the photographs in this text, the stable hand is marked with an "**X**").

When testing end-feel, slightly increase the contact pressure in your stable hand, and if necessary the forearm of your stable hand, to *fixate* one joint partner. *Stabilize* neighboring joints by increasing the contact area of your grip. With adequate fixation, an end-range test technique can be used as a specific Grade III mobilization.

Fixation is an important component of specific Grade III stretch mobilization techniques, which are performed slowly and sustained for longer periods of time. Fixation can also be supplemented with wedges, belts, and other external fixating devices. External fixating devices are usually not necessary for specific movement testing because these tests use small movements with little force.

Therapist's moving hand

With smaller joints, your mobilizing hand grips the joint partner to be moved as close to the joint space as possible. With larger joints, both your hands and body may move together to apply the movement while fixation is provided by a strap or wedge.

Your moving hand performs the testing or treatment procedure. Your moving hand and fingers should be placed as specifically as possible, close to the joint space, so that the movement occurs specifically at the targeted joint.

Procedure

Joint pre-positioning

For the best effect and to avoid pain, carefully pre-position the joint prior to applying a test or treatment procedure. A uniaxial joint can be pre-positioned within one plane of movement; a biaxial joint in two planes; and a triaxial joint in three planes.

If the intent of the technique is pain relief or relaxation, begin treatment in the actual resting position. As the condition tolerates, re-position the joint nearer to the resting position.

If the intent of the technique is stretching, the joint can be positioned three-dimensionally anywhere within the available range-of-motion. Begin in the resting position and progress toward the restriction outside the resting position. The closer the joint position to the limit of movement, the more effective – and risky – the technique.

Pre-positioning cannot be based solely on established norms or typical movement patterns, as actual patient joint characteristic can vary widely.

Mobilization technique

Apply mobilization techniques slowly so that the patient may interrupt treatment at any time. For best effect, vary the speed and rhythm of the test movement or mobilization to control pain and encourage relaxation.

» **For joint play testing including end-feel** (Grade I - III), move slowly and ease into the Grade III range;

» **For pain relief** (Grade I - IISZ), use oscillations or slow, repetitive, intermittent traction movements, staying well short of the Transition Zone;

» **For relaxation** (Grade I - IITZ), apply slow intermittent traction mobilizations, staying well short of the First Stop;

» **For stretching** (Grade III), apply linear traction or glide movements even more slowly and sustain each stretch for 30 - 40 seconds or more. For the longest lasting effect, repeat the stretch in a cyclic manner for a 10 - 15 minute session or to patient tolerance. Note that home exercise is usually necessary to maintain the mobility gains.

Use sound ergonomic principles. When treating larger joints, position yourself close to the patient with your feet apart to maintain a solid base of support. Use gravity and your body weight to generate forces if necessary.

A common error of novice manual therapists is to stand still and use only their hands and arms to mobilize a joint. Produce and control movement not only through your hand movement but also through your body movement. Novice manual therapists must practice and perfect their own body movements before they can accurately evaluate and effectively treat with specific manual therapy techniques.

Therapist safety and treatment effectiveness are further enhanced by:

» Diligent use of body mechanics to protect your body from the rigors of long hours of manual therapy practice (e.g., by absorbing movement forces through your legs rather than through your back).

» Adjustable treatment tables, fixation belts, sand bags, wedges, and other ergonomic and patient positioning aids. Such assistive devices are frequently used in our system.

» Allowing the patient to assist a "passive" movement actively. This lessens the effort exerted by the manual therapist to produce and control a particular movement, but can only be used if the patient can assist without creating muscular tension at the joint targeted for evaluation or treatment.

■ Mobilization Progressions

Treatment progressions for Grade III stretch-mobilizations are illustrated with pre-positioning at a theoretical limit of joint motion. As range-of-motion improves, the joint can be positioned further into the new range.

Symbols

In the photographs which describe each technique in this book, we use the following symbols:

X = Fixation

⟶ = Direction of linear movement (testing and treatment)

CHAPTER 8

FINGERS

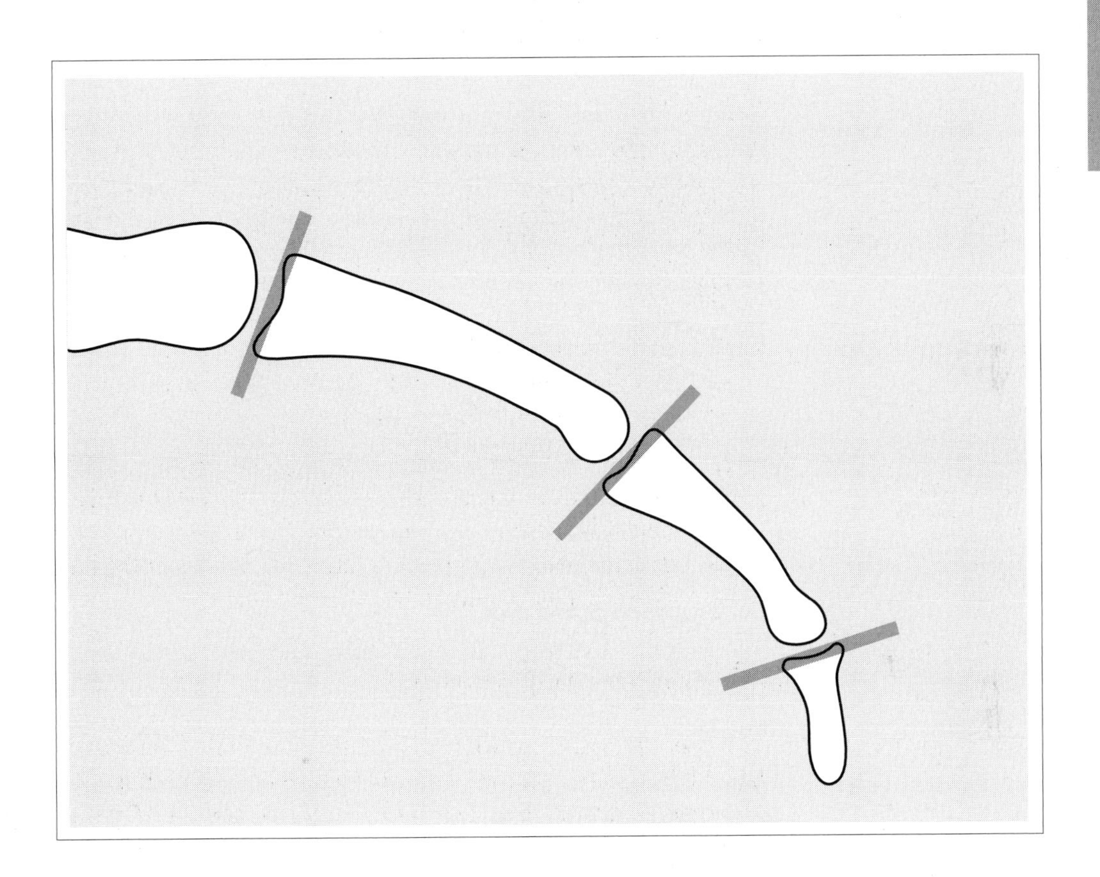

8 Fingers

Functional anatomy and movement

Finger joints

(artt. interphalangeales manus distalis et proximalis, abbreviated DIP and PIP)

The finger joints are anatomically and mechanically simple uniaxial hinge joints (ginglymus, modified sellar). Each phalanx has a head or distal end with a convex surface, a body, and a base or proximal end with a concave surface. The trochlea of the head of the phalanx has a sulcus. The eminence on the base of the phalanx fits into the guiding sulcus provided by the head.

Bony palpation

- Finger bones
- DIP and PIP joint spaces
- One sesamoid bone on IP I

Ligaments

- Collateral ligaments
- Palmar ligaments

Bone movement and axes

- Flexion - extension: around a transverse (radial-ulnar) axis through the head of the phalanx

End feel

- Firm

Joint movement (gliding)

- Concave Rule

Treatment plane

- On the concave joint surface at the base of the phalanx

Zero position

- The longitudinal axis through the metacarpal and corresponding phalangeal bone forms a straight line

Resting position

- Slight flexion in all joints

Close-packed position

- DIPs, PIPs and MCP I: maximal extension

Capsular pattern

- Restricted in all directions with slightly more limitation into flexion

■ "Knuckle" joints II-V

(artt. metacarpophalangeales, abbreviated MCP)

The "knuckle" joints of the 2nd through 5th digits are anatomically and mechanically simple biaxial joints (ellipsoid, modified ovoid). The convex surface of the head, on the distal part of the metacarpal bone, fits into the concave surface of the base of the proximal phalanx. When the hand is fisted, the MCP joints lie approximately one centimeter distal to the knuckles. There are also guiding sulci on the heads of the metacarpals; when the fingers are individually flexed, each finger tip moves towards the middle of the hand.

Bony palpation

- Proximal phalanges II-V
- Metacarpal bones II-V
- Joint spaces of MCP joints II-V
- One sesamoid bone on MCP II and MCP V

Ligaments

- Collateral ligaments
- Palmar ligaments

Bone movement and axes

- Flexion - extension: around a transverse (radial-ulnar) axis through the head of the metacarpal bone
- Radial flexion - ulnar flexion: around a sagittal (dorsal-palmar) axis through the head of the metacarpal bone. Alternate terminology: Abduction and adduction of the fingers is defined as movement away from and toward the third digit. Movement of the middle finger around the dorsal-palmar axis is called radial and ulnar deviation.
- Passive rotation: around a longitudinal axis through the phalanx

End feel

- Firm

Joint movement (gliding)

- Concave Rule

Treatment plane

- On the concave joint surface at the base of the proximal phalanx

Zero position

- The longitudinal axis through the metacarpal and corresponding phalangeal bone forms a straight line

Resting position

- Slight flexion and ulnar flexion

Close-packed position

- Maximal flexion

Capsular pattern

- Restricted in all directions with slightly more limitation into flexion

■ Metacarpal-phalangeal joint of the thumb

(art. metacarpophalangealis I, abbreviated MCP I)

The metacarpal-phalangeal joint of the thumb is an anatomically and mechanically simple uniaxial joint (ginglymus, modified sellar) with a very lax capsule.

Bony palpation

- Proximal phalanx I
- Joint space of MCP I
- 2 sesamoid bones of MCP I

Ligaments

- Collateral ligaments

Bone movement and axes

- Flexion - extension: around a transverse (radial-ulnar) axis through the head of metacarpal I

End feel

- Firm

Joint movement (gliding)

- Concave Rule: The concave surface is on the proximal end of the phalangeal bone; the convex surface is on the distal end of the metacarpal bone.

Treatment plane

- On the concave joint surface at the base of phalanx I

Zero position

- The longitudinal axis through the metacarpal and corresponding phalangeal bone forms a straight line.

Resting position

- Slight flexion

Close-packed position

- Maximal extension

Capsular pattern

- Restricted in all directions with slightly more limitation into flexion

Finger examination scheme

(Refer to Chapters 3 and 4 for more information on examination)

Tests of function

1. Active and passive movements, including stability tests and end-feel

Flexion	*DIP*	45°- 60°
	PIP	100°
	MCP I-V	90°
Extension from zero	*MCP II-V*	10°- 30°
Abduction	*MCP II-V*	total of 90°

2. Translatoric joint play movements, including end-feel

Traction - compression	(Figure 1a)
Gliding	
Palmar	(Figure 2a)
Dorsal	(Figure 2d)
Radial	(Figure 3a)
Ulnar	(Figure 3c)

3. Resisted movements

	ACTS ON:
Flexion	
Flexor digitorum superficialis	PIP
Flexor digitorum profundus	DIP
Lumbricals	MCP
Flexor pollicis brevis	MCP
Flexor pollicis longus	IP
Extension	
Lumbricals	DIP, PIP
Extensor digitorum	DIP, PIP
Extensor digiti minimi	DIP, PIP
Extensor indicis	DIP, PIP
Extensor pollicis brevis	MCP
Extensor pollicis longus	IP
Abduction	
Dorsal interossei	MCP
Abductor digiti minimi	MCP
Adduction	
Palmar interossei	MCP

4. Passive soft tissue movements

Physiological

Accessory

5. Additional tests

Trial treatment

Traction (Figure 1b)

Fingers

■ Finger techniques

Finger traction

for pain and hypomobility

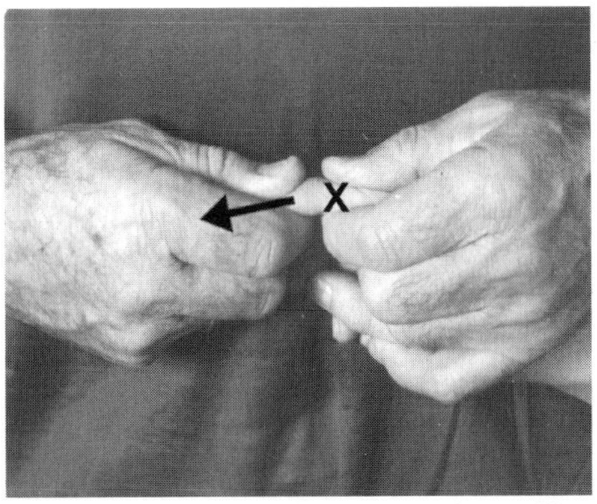

Figure 1a – test and mobilization in resting position

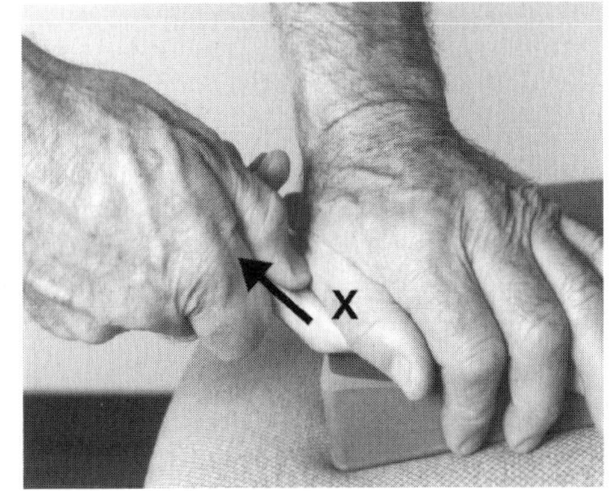

Figure 1b – mobilization in resting position

■ Figure 1a: Test and mobilization in resting position

Objective

- To evaluate the quantity and quality of traction joint play in a DIP, PIP, or MCP joint, including end-feel.
- To decrease pain or increase range-of-motion in a DIP, PIP, or MCP joint.

Starting position

- The patient's palm faces down.
- Position the joint in its resting position.

Hand placement and fixation

- **Therapist's stable hand (left):** Hold the patient's hand and finger in your hand; fixate the patient's hand against your body; grip with your fingers just proximal to the targeted joint space.
- **Therapist's moving hand (right):** Hold the patient's finger in your hand; grip with your fingers just distal to the targeted joint space.

Procedure

- Apply a Grade I, II, or III distal traction movement to the distal phalanx.

■ Figure 1b: Alternate mobilization technique in resting position

- Traction the MCP joint with the dorsal side of the patient's hand resting on a wedge: fixate the patient's metacarpal bone against the wedge with your hand; grip with your thenar eminence just proximal to the patient's MCP joint space; apply a Grade III distal traction movement.

Finger traction
for restricted flexion and extension

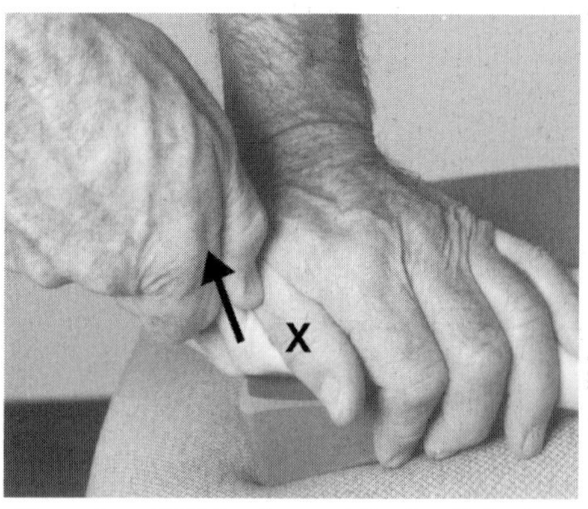

Figure 1c – MCP traction-mobilization in flexion

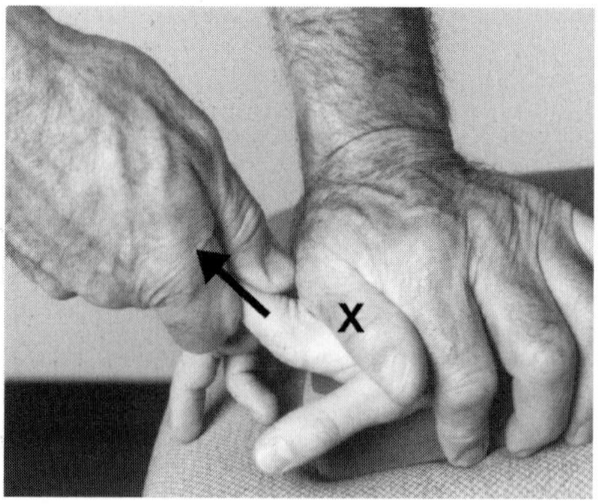

Figure 1d – MCP traction-mobilization in extension

■ Figure 1c: Flexion progression

Objective

- To increase flexion range-of-motion in a DIP, PIP, or MCP joint.

Starting position

- The dorsal side of the patient's hand rests on a wedge.
- Position the joint close to its end range-of-motion in flexion.

Hand placement and fixation

- **Therapist's stable hand (left):** Fixate the patient's proximal joint partner against the wedge with your hand; grip with your thenar eminence just proximal to the targeted joint space.
- **Therapist's moving hand (right):** Hold the patient's finger in your hand; grip with your fingers just distal to the targeted joint space.

Procedure

- Apply a Grade III distal traction movement to the distal phalanx.

■ Figure 1d: Extension progression for the MCP joint

- The patient's palm rests on a wedge with the MCP joint positioned close to its end range-of-motion in extension. Apply a Grade III distal traction movement to the distal phalanx.

Finger palmar glide
for restricted flexion

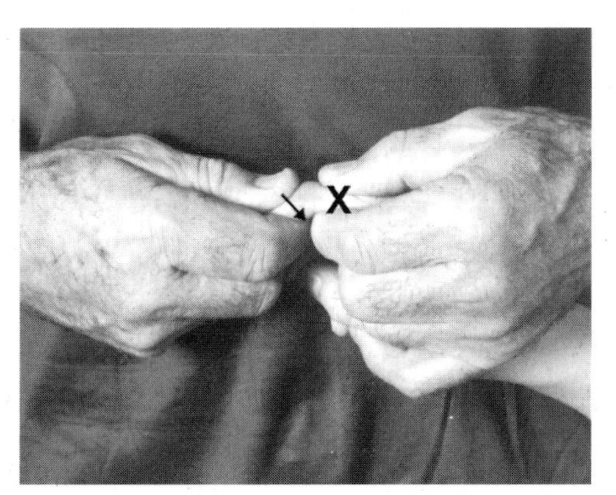

Figure 2a – test and mobilization in resting position

Fingers

■ Figure 2a: Test and mobilization in resting position

Objective

- To evaluate the quantity and quality of palmar glide joint play in a DIP, PIP, or MCP joint, including end-feel.
- To increase flexion range-of-motion in a DIP, PIP, or MCP joint (Concave Rule).

Starting position

- The patient's palm faces down.
- Position the joint in its resting position.

Hand placement and fixation

- **Therapist's stable hand (left):** Hold the patient's hand and finger in your hand; fixate the patient's hand against your body; grip with your fingers just proximal to the targeted joint space.
- **Therapist's moving hand (right):** Hold the patient's finger in your hand; grip with your fingers just distal to the targeted joint space.

Procedure

- Apply a Grade II or III palmar glide movement to the distal phalanx.

Fingers

Finger palmar glide
for restricted flexion (cont'd)

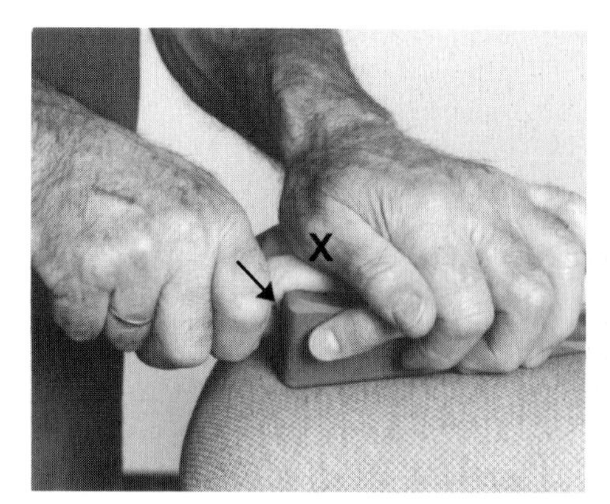

Figure 2b – mobilization in resting position

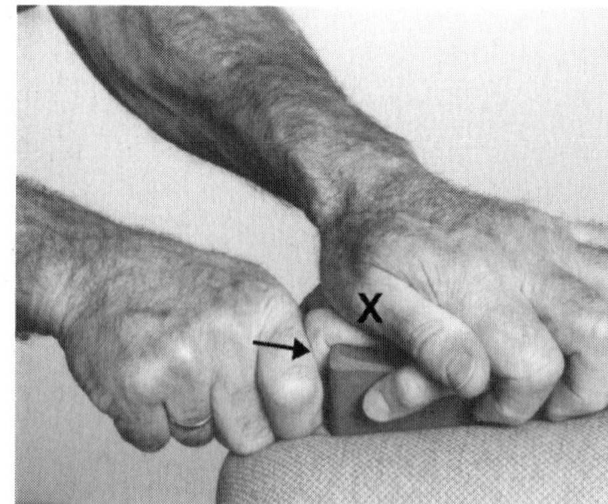

Figure 2c – mobilization in flexion

■ Figure 2b: Alternate mobilization technique in resting position

Objective

- To increase flexion range-of-motion in a DIP, PIP, or MCP joint (Concave Rule).

Starting position

- The patient's palm rests on a wedge.
- Position the joint in its resting position.

Hand placement and fixation

- **Therapist's stable hand (left):** Fixate the patient's proximal joint partner against the wedge with your hand; grip with your thenar eminence just proximal to the targeted joint space.
- **Therapist's moving hand (right):** Hold the patient's finger in your hand; grip with your fingers just distal to the targeted joint space.

Procedure

- Apply a Grade III palmar glide movement to the distal phalanx.

■ Figure 2c: Flexion progression

- Apply a Grade III palmar glide movement with the targeted finger joint positioned close to its end range-of-motion in flexion.

Finger dorsal glide
for restricted extension

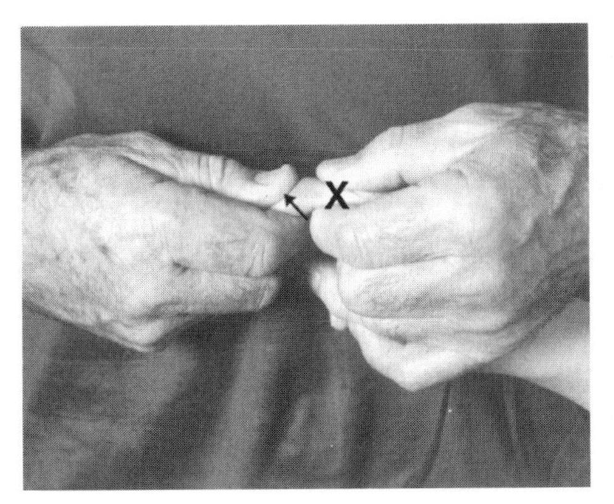

Figure 2d – test and mobilization in resting position

Fingers

■ Figure 2d: Test and mobilization in resting position

Objective

- To evaluate the quantity and quality of dorsal glide joint play in a DIP, PIP, or MCP joint, including end-feel.
- To increase extension range-of-motion in a DIP, PIP, or MCP joint (Concave Rule).

Starting position

- The patient's palm faces down.
- Position the joint in its resting position.

Hand placement and fixation

- **Therapist's stable hand (left):** Hold the patient's hand and finger in your hand; fixate the patient's hand against your body; grip with your fingers just proximal to the targeted joint space.
- **Therapist's moving hand (right):** Hold the patient's finger in your hand; grip with your fingers just distal to the targeted joint space.

Procedure

- Apply a Grade II or III dorsal glide movement to the distal phalanx.

Finger dorsal glide
for restricted extension (cont'd)

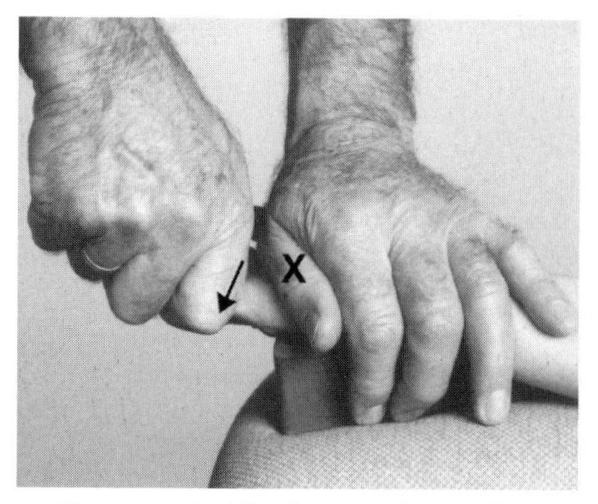

Figure 2e – mobilization in resting position

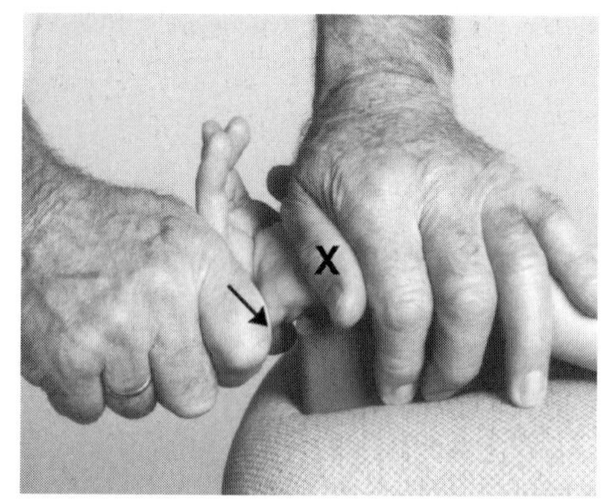

Figure 2f - mobilization in extension

■ Figure 2e: Mobilization in resting position

Objective

- To increase extension range-of-motion in a DIP, PIP, or MCP joint (Concave Rule).

Starting position

- The dorsal side of the patient's hand rests on a wedge.
- Position the joint in its resting position.

Hand placement and fixation

- **Therapist's stable hand (left):** Fixate the patient's proximal joint partner against the wedge with your hand; grip with your thenar eminence just proximal to the targeted joint space.
- **Therapist's moving hand (right):** Hold the patient's finger in your hand; grip with your fingers just distal to the targeted joint space.

Procedure

- Apply a Grade III dorsal glide movement to the distal phalanx.

■ Figure 2f: Extension progression

- Apply a Grade III dorsal glide movement with the targeted finger joint positioned close to its end range-of-motion in extension.

Finger radial glide
for restricted flexion and extension

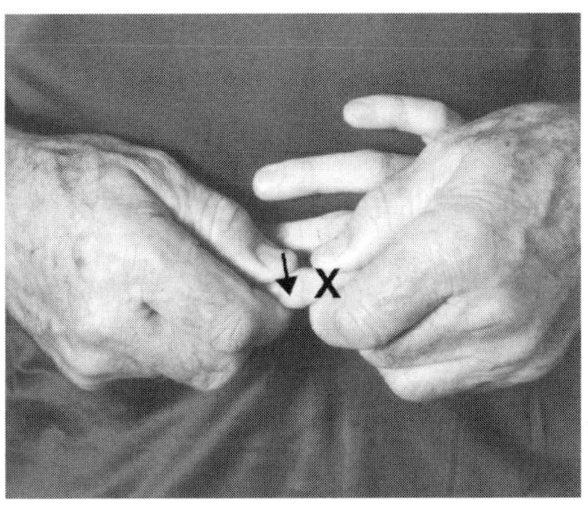

Figure 3a – test and mobilization in resting position

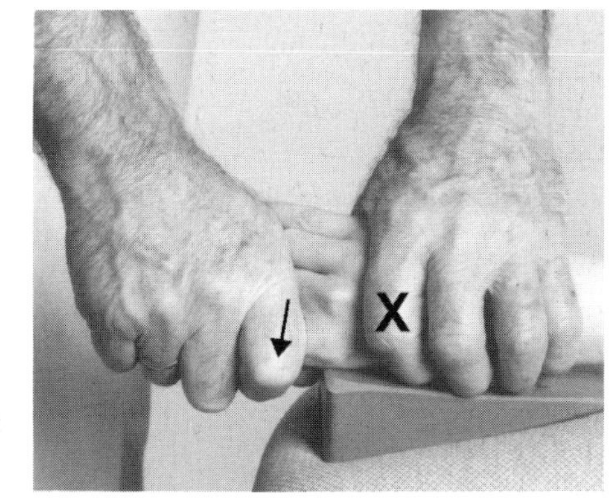

Figure 3b - mobilization in resting position

■ Figure 3a: Test and mobilization in resting position

Objective

- To evaluate the quantity and quality of radial glide joint play in a DIP, PIP, or MCP joint, including end-feel.
- To increase flexion and extension range-of-motion in a DIP, PIP, or MCP joint (Concave Rule).

Starting position

- The patient's palm faces the therapist.
- Position the joint in its resting position.

Hand placement and fixation

- **Therapist's stable hand (left):** Hold the patient's hand and finger in your hand; fixate the patient's hand against your body; grip with your fingers just proximal to the targeted joint space.
- **Therapist's moving hand (right):** Hold the patient's finger in your hand; grip with your fingers just distal to the targeted joint space.

Procedure

- Apply a Grade II or III radial glide movement to the distal phalanx.

■ Figure 3b: Alternate mobilization technique in resting position

- Apply a Grade III radial glide movement to the MCP joint with the radial side of the patient's hand resting on a wedge. Fixate the patient's metacarpal bone against the wedge with your hand; grip just proximal to the patient's MCP joint space.
- This technique can be used to increase radial glide joint play (Concave Rule) both in the resting position and approaching the end range-of-motion into flexion, extension, and MCP radial flexion.

Finger ulnar glide

for restricted flexion and extension

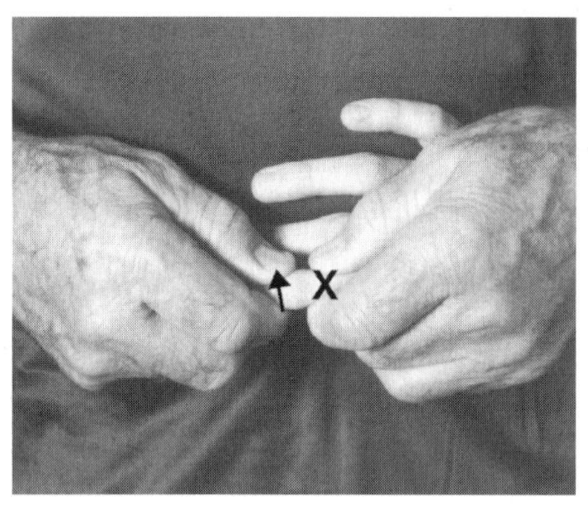

Figure 3c – test and mobilization in resting position

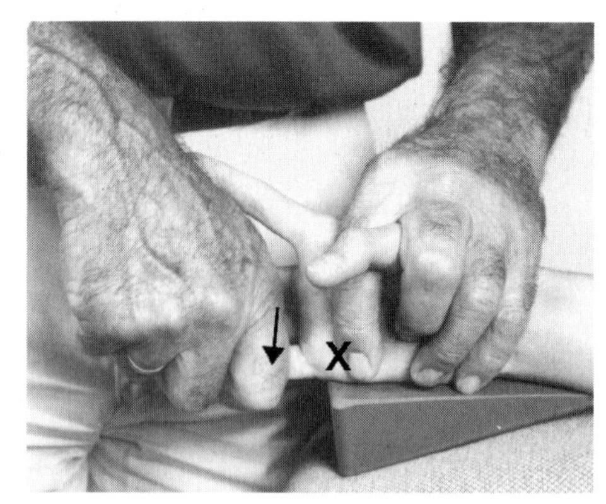

Figure 3d - mobilization in resting position

■ Figure 3c: Test and mobilization in resting position

Objective

- To evaluate the quantity and quality of ulnar glide joint play in a DIP, PIP, or MCP joint, including end-feel.
- To increase flexion and extension range-of-motion in a DIP, PIP, or MCP joint (Concave Rule).

Starting position

- The patient's palm faces the therapist.
- Position the joint in its resting position.

Hand placement and fixation

- **Therapist's stable hand (left):** Hold the patient's hand and finger in your hand; fixate the patient's hand against your body; grip with your fingers just proximal to the targeted joint space.
- **Therapist's moving hand (right):** Hold the patient's finger in your hand; grip with your fingers just distal to the targeted joint space.

Procedure

- Apply a Grade II or III ulnar glide movement to the distal phalanx.

■ Figure 3d: Alternate mobilization technique in resting position

- Apply a Grade III ulnar glide movement in an MCP joint with the ulnar side of the patient's hand resting on a wedge; fixate the patient's metacarpal bone against the wedge with your hand; grip just proximal to the patient's MCP joint space.
- This technique can be used to increase ulnar glide joint play (Concave Rule) both in the resting position and near the end range-of-motion into flexion, extension, or MCP ulnar flexion.

CHAPTER 9

METACARPALS

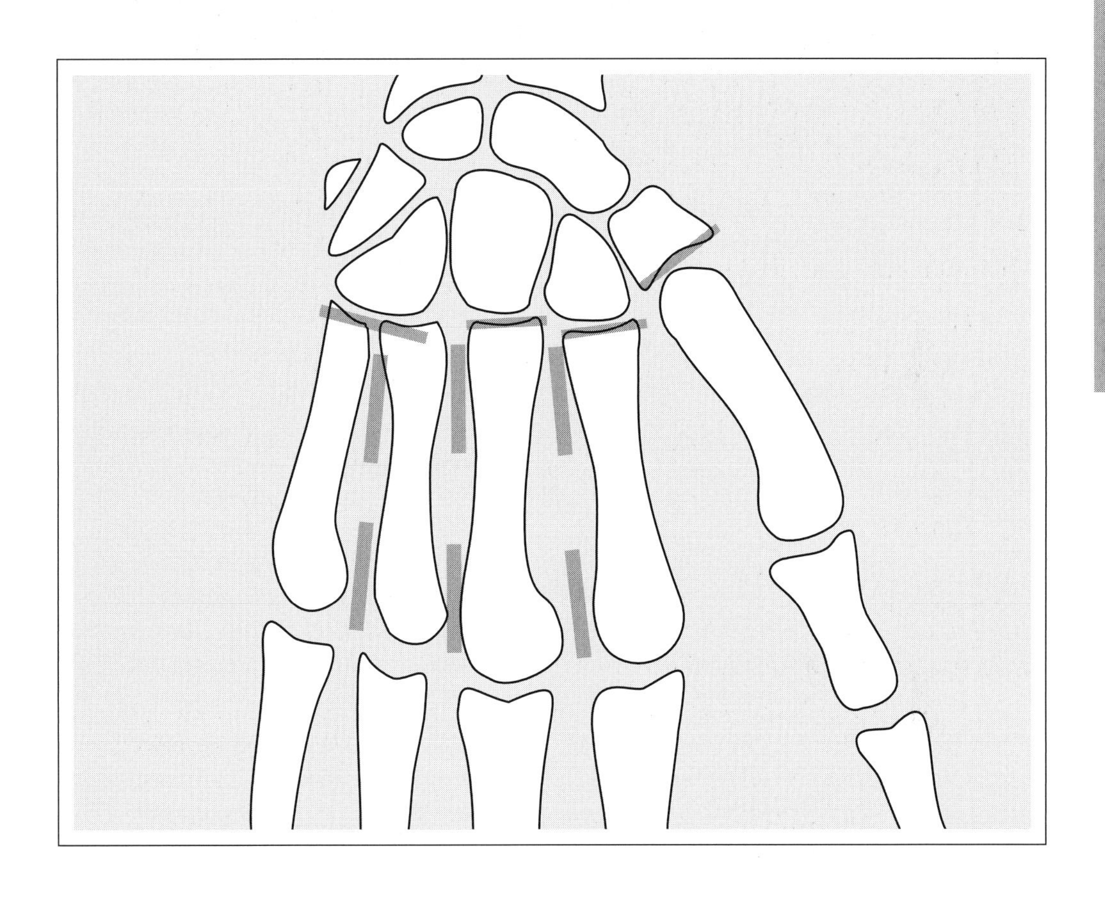

9 Metacarpals

Functional anatomy and movement

Metacarpals

When there is movement restriction in the hand, first treat the metacarpals and later progress to treatment of the fingers.

Hand proper
(metacarpus)

The hand proper is made up of five metacarpal bones corresponding to the five digits. Each metacarpal has a head or distal end with a convex surface, a body, and a base or proximal end with a concave surface. The heads of the metacarpals have no joints in relation to each other, but are joined together by the deep transverse metacarpal ligaments.

The joints between the bases of the second through fifth metacarpals and the adjacent row of carpal bones (*artt. carpometacarpales*; abbreviated CM), and the joints between the bases of the metacarpal bones (*artt. intermetacarpales*; abbreviated IM) are plane or nearly flat. However, these joints individually are anatomically simple and mechanically compound plane amphiarthroses (Figure 13). All of these "plane" joints have a small curvature, but this need not be taken into consideration during treatment, since only traction and dorsal-palmar gliding techniques are used.

The intermetacarpal joints of the hand share one complex cavity with the carpometacarpal joints I-V. Therefore, all these joints together are often called the "big carpometacarpal joint." This complex joint cavity does not communicate with the first carpometacarpal or pisiform joints.

The dorsal convex arch of the hand proper changes shape with all finger movements.

Bony palpation

- Metacarpals II-V
- Distal row of carpals (trapezoid, capitate, hamate)
- Carpometacarpal joint space II-V

Ligaments

- Dorsal interosseous ligaments
- Palmar metacarpal ligaments
- Dorsal and palmar carpometacarpal ligaments

Bone movement and axes

There is relatively more movement in the ulnar metacarpal joints than in the radial metacarpal joints. For example, the metacarpal V - hamate articulation, a saddle joint, is capable of flexion-extension, radial-ulnar flexion, and also opposition.

Intermetacarpal joints:

- There are no defined axes for the small movements that occur in these joints. As the curve of the transverse metacarpal arch increases, the metacarpals move in a palmar direction with relation to metacarpal III.
- As the curve of the transverse metacarpal arch decreases, the metacarpals move in a dorsal direction with relation to metacarpal III.

Carpometacarpal joints:

- Flexion - extension: around a transverse (radial-ulnar) axis through the carpal bones
- Radial - ulnar flexion: around a sagittal (dorsal-palmar) axis through the carpal bones

End feel

- Firm

Joint movement (gliding)

- Concave Rule

Treatment plane

- Distal and proximal intermetacarpal II-V treatment plane lies between and perpendicular to the metacarpal bones
- Carpometacarpal treatment plane lies on the concave joint surface at the base of the metacarpal

Zero position

- *CM joints II-V:* not described

Resting position

- *CM joints II-V:* not described

Close-packed position

- Unknown

Capsular pattern

- *CM joints II-V:* limited equally in all directions

Metacarpals

■ First ("little") carpometacarpal joint
(art. carpometacarpalis pollicis)

The first carpometacarpal joint between the first metacarpal bone and trapezium, is an anatomically and mechanically simple biaxial joint (sellaris, unmodified sellar). It must be treated as a saddle joint, but because there is a lax capsule it is functionally a triaxial ball and socket joint (sphaeroidea).

Bony palpation

- Base of metacarpal I
- Trapezium
- Carpometacarpal I joint space

Ligaments

- Strengthen the capsule on all sides

Bone movement and axes

The joint surface of the trapezium is not parallel to the joint surfaces of the other distal carpal bones because the trapezium is rotated 90° towards the palm. Therefore, when describing movements of the first carpometacarpal joint it must be remembered that the axes are also rotated 90°.

- Palmar abduction - adduction: The base of the first metacarpal bone moves with its convex surface around a radial-ulnar axis through its base.
- Flexion - extension: The base of the first metacarpal bone moves with its concave surface around a dorsal-palmar axis through the trapezium.
- Rotation: The axis passes longitudinally through the metacarpal bone. Rotation can only be performed passively.
- Opposition - reposition: Opposition occurs when the abducted thumb is flexed; reposition occurs when the adducted thumb is extended.

End feel

- Firm

Joint movement (gliding)

- Flexion - extension: Concave Rule
- Abduction - adduction: Convex Rule

Treatment plane

- Flexion - extension: on the concave joint surface at the base of the metacarpal.
- Abduction - adduction: on the concave joint surface of the trapezium

Zero position

- MC I bone midway between maximal abduction-adduction and flexion-extension from zero.

Resting position

- MC I bone midway between abduction-adduction and flexion-extension

Close-packed position

- Maximal opposition

Capsular pattern

- Abduction-extension

Metacarpal examination scheme

(Refer to Chapters 3 and 4 for more information on examination)

Tests of function

1. Active and passive movements, including stability tests and end-feel

CM I	Flexion - extension	50° total
	Abduction - adduction	40° total
CM II-V	Flexion - extension	little movement
	Abduction - adduction	
CM V	Opposition	

2. Translatoric joint play, including end-feel

CM I	Traction - compression	(Figure 7a)
	Gliding	
	Ulnar	(Figure 7c)
	Radial	(Figure 7e)
	Palmar	(Figure 7g)
	Dorsal	(Figure 7i)
CM II-V	Traction - compression	(Figure 6a)
	Gliding	
	Palmar	
	Dorsal	
IMC II-V	Gliding	(Figure 5a)

3. Resisted movements

			ACTS ON:
CM I	Flexion:	*Flexor pollicis longus*	IP, MCP
		Flexor pollicis brevis	IP
	Extension:	*Extensor pollicis longus*	IP, MCP
		Extensor pollicis brevis	IP
	Abduction:	*Abductor pollicis longus*	
		Abductor pollicis brevis	
	Adduction:	*Adductor pollicis*	
	Opposition:	*Opponens pollicis,*	
		Flexor pollicis brevis	
CM V	Opposition:	*Opponens digiti minimi*	

4. Passive soft tissue movements

Physiological

Accessory

5. Additional tests

Trial treatment

CM I	Traction	(Figure 7b)
CM II-V	Traction	(Figure 6b)

Metacarpal techniques

Metacarpals

Recommended mobilization sequence for the hand

1. Carpometacarpal traction (Figure 6)
2. Proximal metacarpal glide (Figure 5)
3. Metacarpal arch mobilization (Figure 4)

Metacarpal arch
for hypomobility

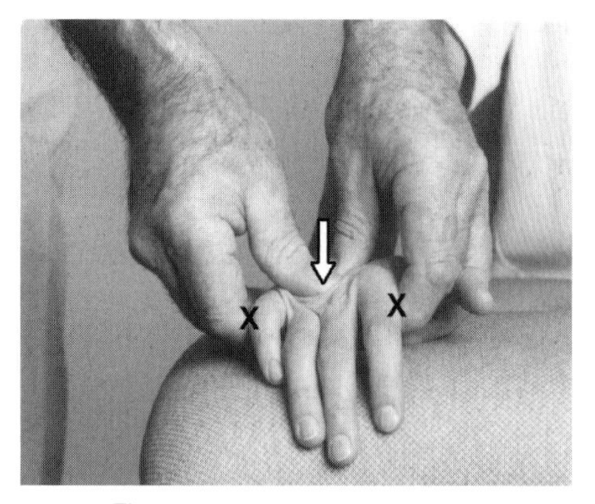

Figure 4a – test and mobilization

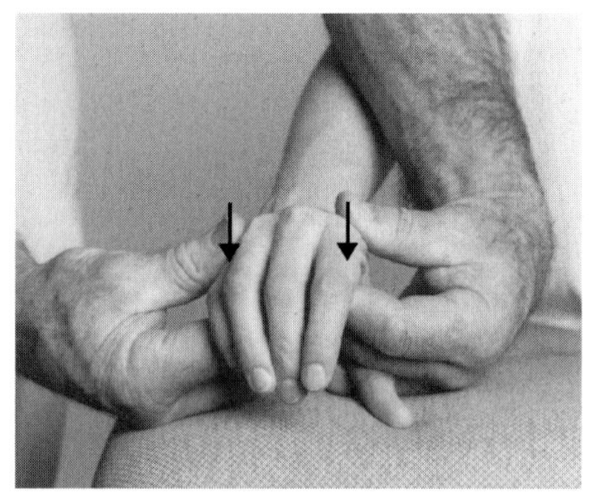
Figure 4b – test and mobilization

■ Figure 4a: Test and mobilization, dorsal-concave arch

Objective

- To evaluate the quantity and quality of metacarpal arch mobility, including end-feel.
- To increase metacarpal arch mobility; to stretch the distal syndesmosis. See Figure 5b for an alternate stretch for the distal syndesmosis.

Starting position

- The patient's palm rests on the treatment surface.

Hand placement and fixation

- Grip the patient's hand with your fingers on the palmar side of the patient's hand to provide fixation; place your thumbs together on the dorsal side of metacarpal III to apply the movement.

Procedure

- Apply a Grade II or III downward pressure with your thumbs to reverse the patient's metacarpal arch (dorsal concave).

■ Figure 4b: Test and mobilization, dorsal-convex arch

- Fixate metacarpal III with your fingers on the palmar surface of the hand; mobilize with your thumbs pressing down on metacarpals II and IV/V.

Proximal intermetacarpal palmar glide
for hypomobility

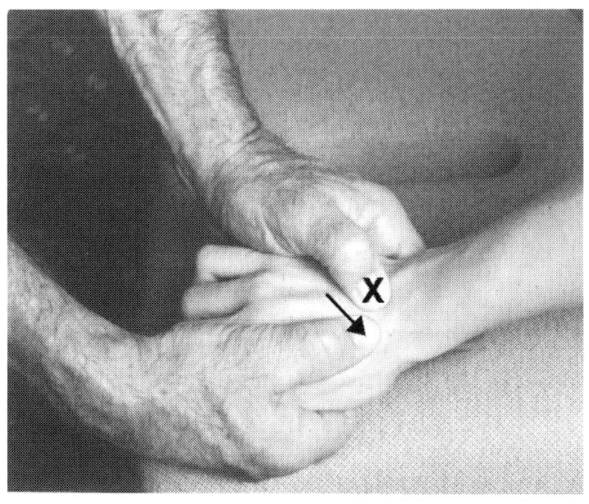

Figure 5a – test and mobilization in resting position

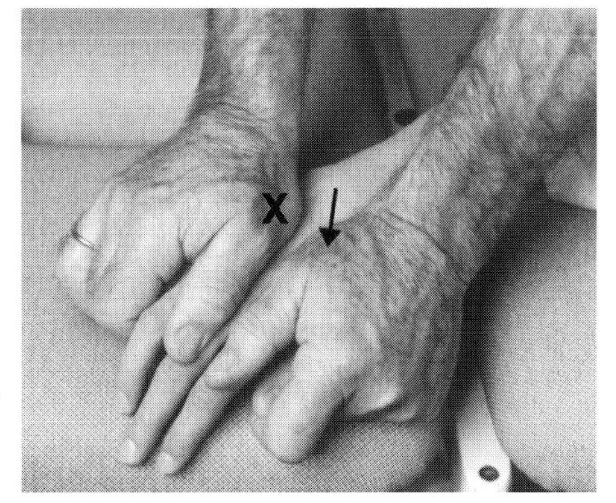

Figure 5b – mobilization in resting position

■ Figure 5a: Test and mobilization in resting position

Objective

- To evaluate the quantity and quality of metacarpal palmar glide joint play, including end-feel.
- To increase intermetacarpal mobility.

Starting position

- The patient's palm rests on the treatment surface.

Hand placement and fixation

- **Therapist's stable hand (left):** Hold the patient's hand from the ulnar side; grip your thumb and fingers around the base of the patient's metacarpal (metacarpal III shown).
- **Therapist's moving hand (right):** Hold the patient's hand from the radial side; grip with your thumb and fingers around the base of the adjacent metacarpal (metacarpal II shown).

Procedure

- Press your right hand downward to apply a Grade II or III palmar glide movement.

■ Figure 5b: Mobilization in resting position

- Grip for fixation and mobilization with your thenar eminences and thumbs. Apply a Grade III palmar glide movement.

■ Alternate technique (not shown)

- Apply a Grade II or III dorsal glide movement to increase intermetacarpal mobility with the dorsal side of the patient's hand facing down.

Carpometacarpal traction
for pain and hypomobility

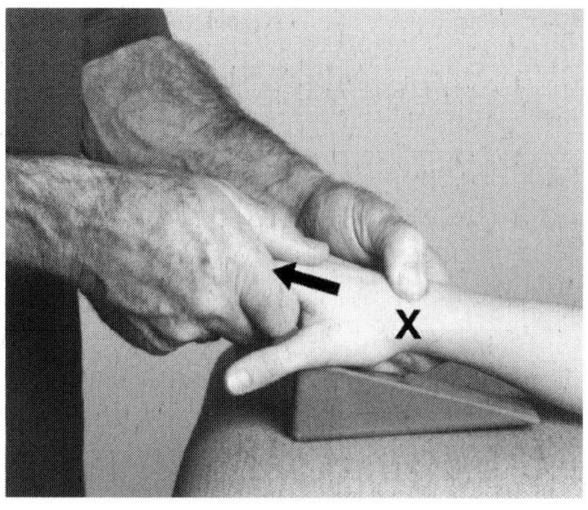

Figure 6a – test and mobilization in resting position

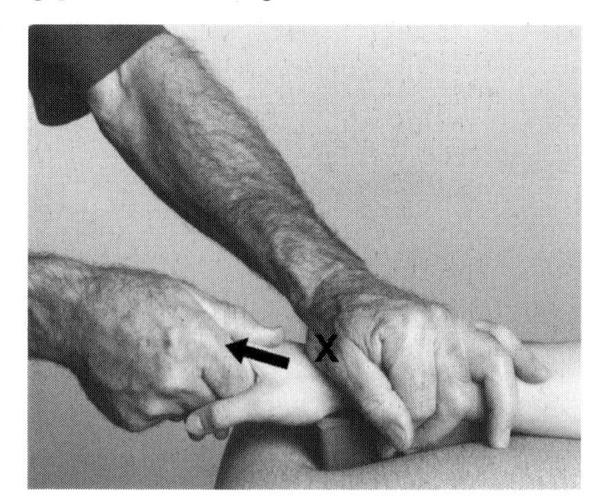

Figure 6b – mobilization in resting position

■ Figure 6a: Test and mobilization in resting position

Objective

- To evaluate the quantity and quality of traction joint play in a carpometacarpal joint, including end-feel.
- To decrease pain or increase range-of-motion in the carpometacarpal joints.

Starting position

- The patient's hand rests on the wedge with their palm facing down.
- Position the joint in its resting position.

Hand placement and fixation

- **Therapist's stable hand (left):** Grip around the targeted carpal bone just proximal to the joint space (trapezii shown).
- **Therapist's moving hand (right):** Grip the patient's targeted metacarpal bone just distal to the joint space (metacarpal II shown).

Procedure

- Apply a Grade I, II or III traction movement to the metacarpal bone; palpate the joint space with your thumb.

■ Figure 6b: Mobilization in resting position

- Apply a Grade III traction movement to increase carpometacarpal mobility. Use your left thenar eminence to fixate the patient's carpal bone against the treatment surface or wedge.

Thumb carpometacarpal traction

for pain and hypomobility

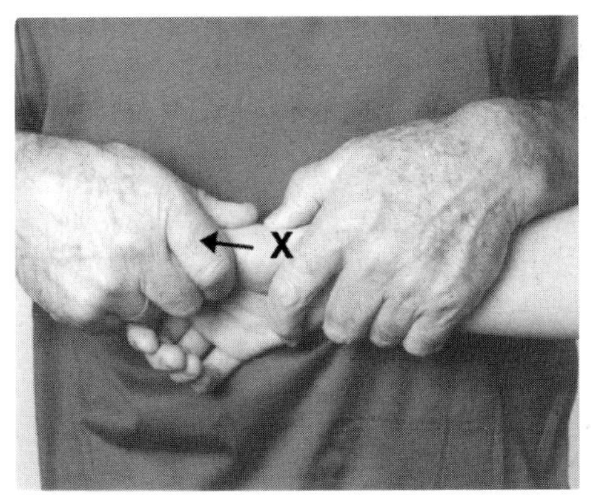

Figure 7a – test and mobilization in resting position

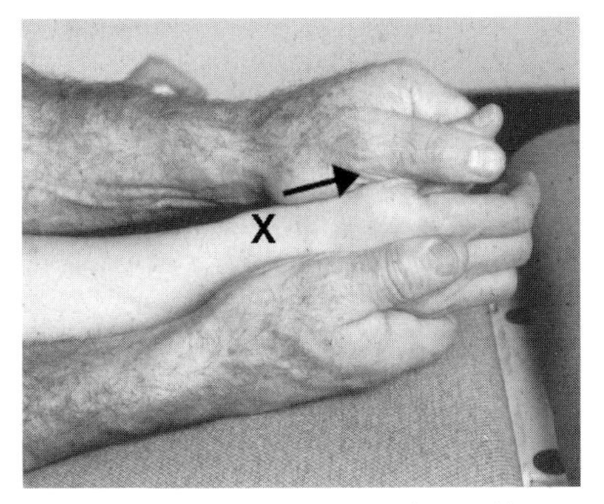

Figure 7b – mobilization in resting position

■ Figure 7a: Test and mobilization in resting position

Objective

- To evaluate the quantity and quality of distal traction joint play in the carpometacarpal I joint, including end-feel.
- To decrease pain or increase range-of-motion in a carpometacarpal joint.

Starting position

- The ulnar side of the patient's hand faces down.
- Position the joint in its resting position.

Hand placement and fixation

- **Therapist's stable hand (left):** Hold the patient's distal forearm with your hand; grip around the trapezium just proximal to the joint space; fixate the patient's hand against your body.
- **Therapist's moving hand (right):** Grip the patient's metacarpal I just distal to the joint space.

Procedure

- Apply a Grade I, II, or III distal traction movement to metacarpal I.

■ Figure 7b: Mobilization in resting position

- Apply a Grade III distal traction movement to increase general mobility of the carpometacarpal I joint. The ulnar side of the patient's hand rests on the treatment surface; your right hand fixates the patient's trapezium; your left hand grips around metacarpal I with your thenar eminence and fingers.

■ Flexion and extension progression (not shown)

- Position the carpometacarpal I joint near the end range-of-motion into flexion or extension.

Thumb metacarpal-carpal ulnar glide
for restricted flexion

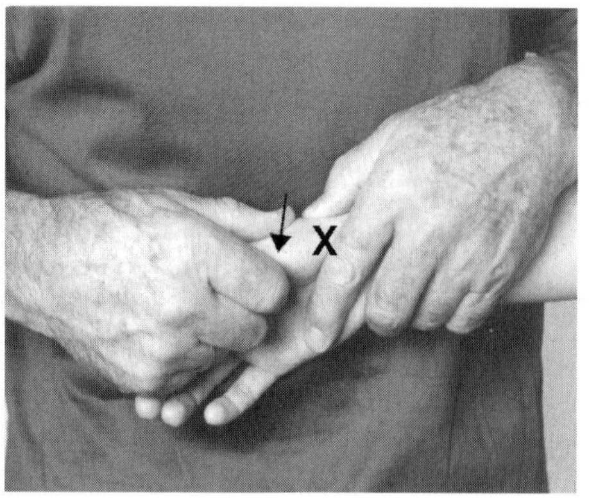

Figure 7c – test and mobilization in resting position

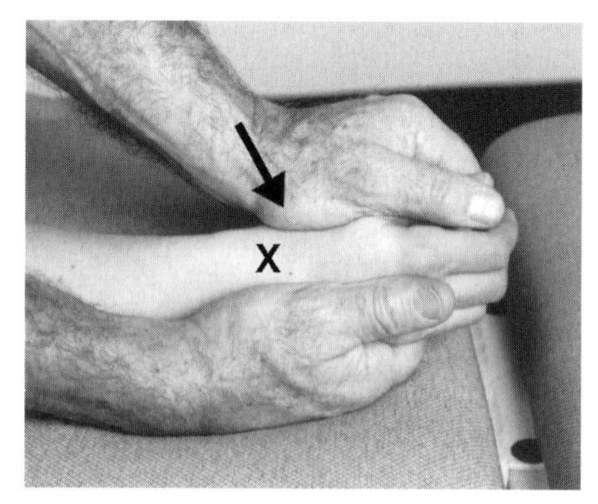

Figure 7d – mobilization in resting position

■ Figure 7c: Test and mobilization in resting position

Objective

- To evaluate the quantity and quality of ulnar glide joint play in the carpometacarpal I joint, including end-feel.
- To increase thumb flexion range-of-motion in the carpometacarpal I joint (Concave Rule).

Starting position

- The ulnar side of the patient's hand faces down.
- Position the joint in its resting position.

Hand placement and fixation

- **Therapist's stable hand (left):** Hold the patient's distal forearm with your hand; grip around the trapezium just proximal to the joint space; fixate the patient's hand against your body.
- **Therapist's moving hand (right):** Grip the patient's metacarpal I just distal to the joint space.

Procedure

- Apply a Grade II or III ulnar glide movement to metacarpal I.

■ Figure 7d: Mobilization in resting position

- Apply a Grade III ulnar glide movement. The ulnar side of the patient's hand rests on the treatment surface; your right hand fixates the patient's trapezium; your left hand grips around metacarpal I with your thenar eminence and fingers.

■ Flexion progression (not shown)

- Position the carpometacarpal I joint near the end range-of-motion into flexion.

Thumb metacarpal-carpal radial glide
for restricted extension

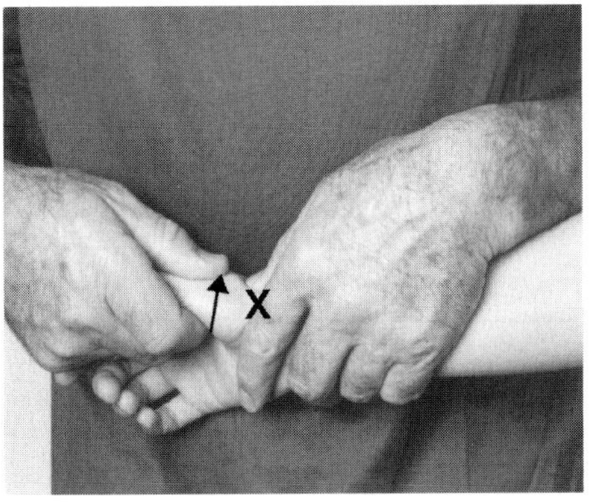

Figure 7e – test and mobilization in resting position

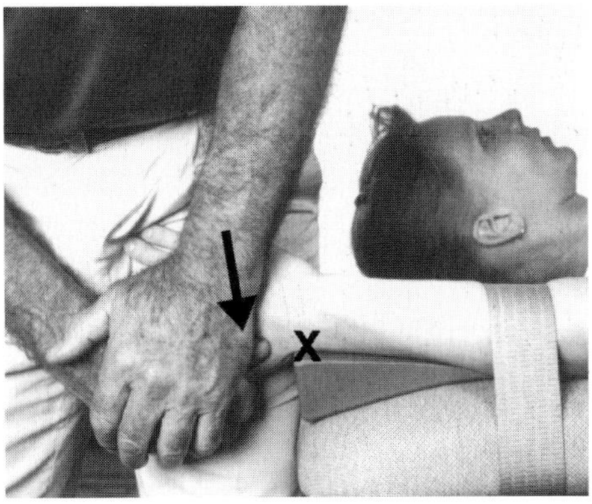

Figure 7f – mobilization in resting position

■ Figure 7e: Test and mobilization in resting position

Objective

- To evaluate the quantity and quality of radial glide joint play in the carpometacarpal I joint, including end-feel.
- To increase thumb extension range-of-motion (Concave Rule).

Starting position

- The ulnar side of the patient's hand faces down.
- Position the joint in its resting position.

Hand placement and fixation

- **Therapist's stable hand (left):** Hold the patient's distal forearm with your hand; grip around the trapezium just proximal to the joint space; fixate the patient's hand against your body.
- **Therapist's moving hand (right):** Grip the patient's metacarpal I just distal to the joint space.

Procedure

- Apply a Grade II or III radial glide movement to metacarpal I.

■ Figure 7f: Mobilization in resting position

- Apply a Grade III radial glide movement to increase thumb extension. The patient lies supine; the radial side of the patient's hand rests on the wedge with the patient's hand extended over the edge; fixate the proximal forearm with a strap; the wedge provides fixation to the distal forearm; grip around metacarpal I with your left thumb and fingers; reinforce your grip with your right hand and lean your body through your extended arm.

■ Extension progression (not shown)

- Position the carpometacarpal I joint near its end range-of-motion into extension.

Thumb metacarpal-carpal palmar glide
for restricted adduction

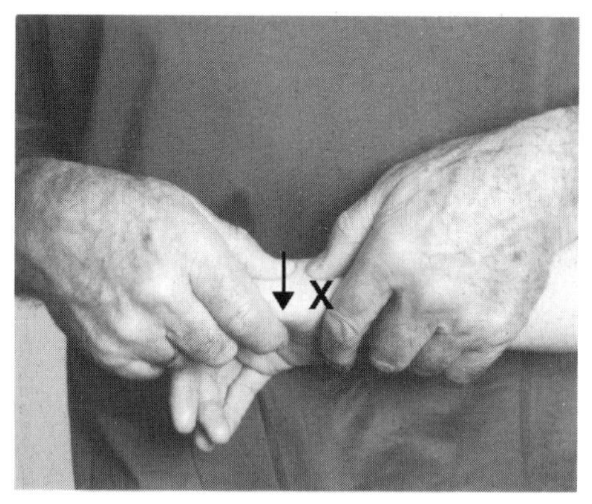

Figure 7g – test and mobilization in resting position

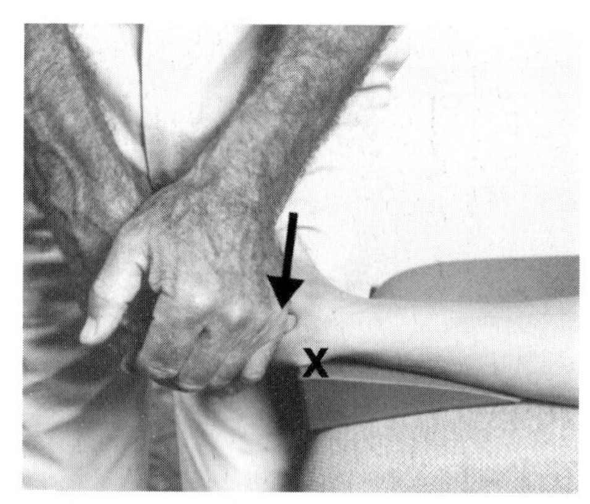

Figure 7h – mobilization in resting position

Metacarpals

■ Figure 7g: Test and mobilization in resting position

Objective

- To evaluate the quantity and quality of palmar glide joint play in the carpometacarpal I joint, including end-feel.
- To increase thumb adduction range-of-motion (Convex Rule).

Starting position

- The palmar side of the patient's hand faces down.
- Position the joint in its resting position.

Hand placement and fixation

- **Therapist's stable hand (left):** Hold the patient's distal forearm with your hand; grip around the trapezium just proximal to the joint space; fixate the patient's hand against your body.
- **Therapist's moving hand (right):** Grip the patient's metacarpal I just distal to the joint space.

Procedure

- Apply a Grade II or III palmar glide movement to metacarpal I.

■ Figure 7h: Mobilization in resting position

- Apply a Grade III palmar glide movement to increase thumb adduction. The palmar side of the patient's hand, including the trapezium, rests on the treatment wedge; grip around metacarpal I with your left thumb and fingers; reinforce your grip with your right hand and lean your body through your extended arm.

■ Adduction progression (not shown)

- Position the carpometacarpal I joint near the end range-of-motion into adduction. Note that in cases of extreme hypomobility the joint may remain in an abducted position.

Thumb metacarpal-carpal dorsal glide
for restricted abduction

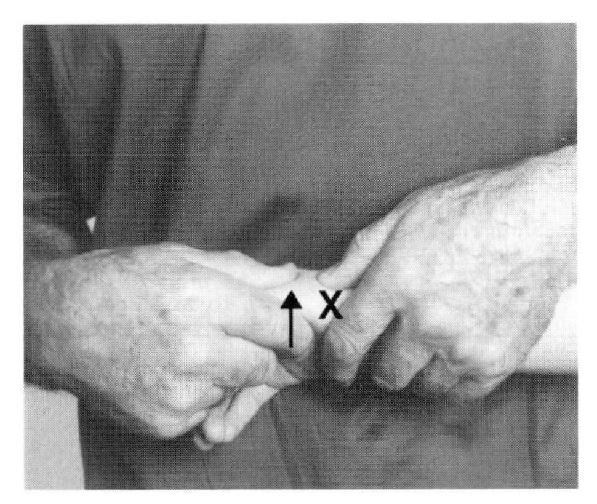

Figure 7i – test and mobilization in resting position

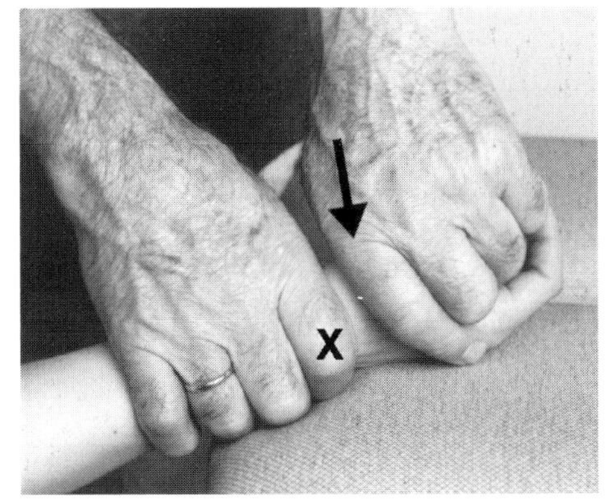

Figure 7j – mobilization in resting position

Metacarpals

■ Figure 7i: Test and mobilization in resting position

Objective

- To evaluate the quantity and quality of dorsal glide joint play in the carpometacarpal I joint, including end-feel.
- To increase thumb abduction range-of-motion (Convex Rule).

Starting position

- The palmar side of the patient's hand faces down.
- Position the joint in its resting position.

Hand placement and fixation

- **Therapist's stable hand (left):** Hold the patient's distal forearm with your hand; grip around the trapezium just proximal to the joint space; fixate the patient's hand against your body.
- **Therapist's moving hand (right):** Grip the patient's metacarpal I just distal to the joint space.

Procedure

- Apply a Grade II or III dorsal glide movement to metacarpal I.

■ Figure 7j: Mobilization in resting position

- Apply a Grade III dorsal glide movement to increase thumb abduction. The dorsal side of the patient's hands rests on the treatment surface; grip around metacarpal I with your left hand; place your metacarpal II-phalangeal joint just distal to the joint space; lean your body through your extended arm.

■ Abduction progression (not shown)

- Position the carpometacarpal I joint near its end range-of-motion into abduction.

■ Notes

Metacarpals

CHAPTER 10

WRIST

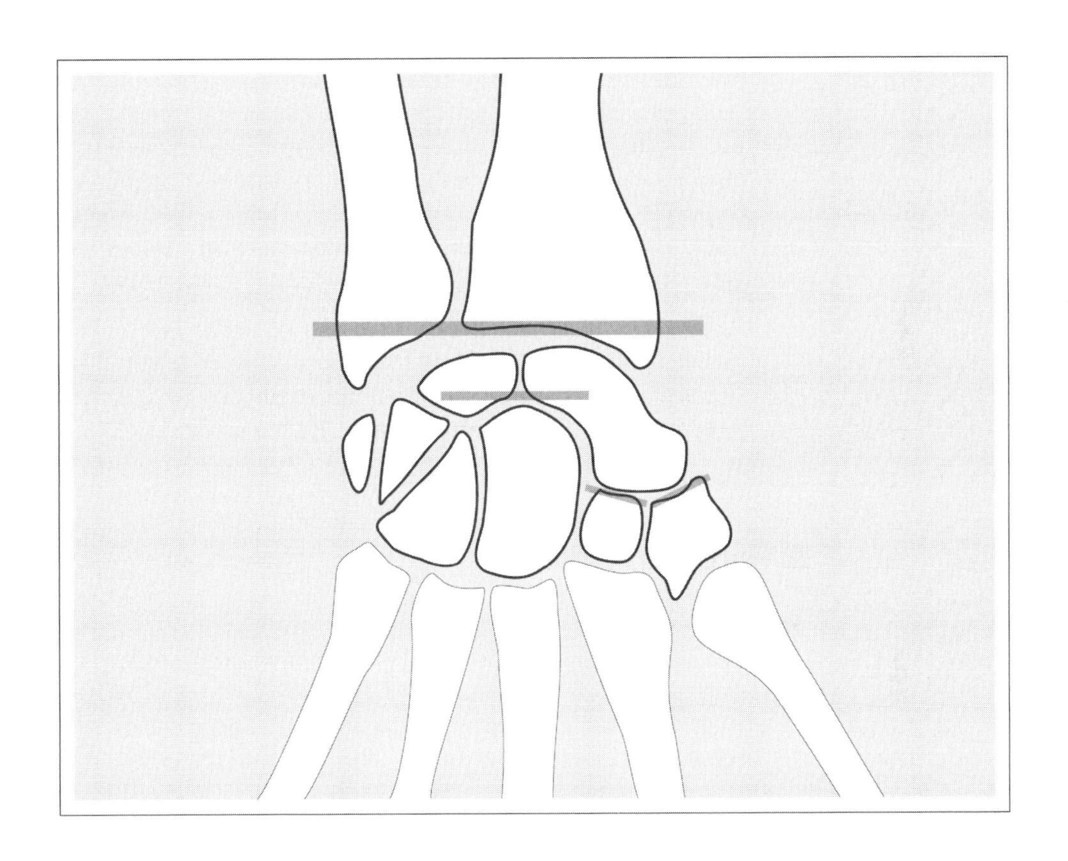

10 Wrist

(carpus)

Functional anatomy and movement

The wrist joint complex consists of eight carpal bones arranged in two rows, the distal radius and ulna, and an articular disc (Figure 13).

The proximal or first row of carpal bones are, starting from the radial side: scaphoid (navicular = os scaphoideum); lunate (semilunate = os lunatum); triquetral (cuneiform = os triquetral); and pisiform (os pisiforme). In the second or distal row, starting from the radial side: trapezium (multangulum majus = os trapezium); trapezoid (multangulum minus = os trapezoideum); capitate (magnum = os capitatum); and hamate (unciform = os hamatum).

An articular disc binds the distal end of the radius and ulna together and its lower surface forms part of the radiocarpal joint. The articular disc takes part in all movements of the radiocarpal joint in addition to forearm pronation and supination.

The wrist is divided into three joints:

Radiocarpal joint

(art. radiocarpalis)

The radiocarpal joint, the "true" wrist joint, is an anatomically and mechanically simple biaxial joint (ellipsoid, modified ovoid). The convex surface is made up of the scaphoid, lunate and triquetral and their interosseous ligaments, which are often calcified. Therefore, these three bones act as one joint surface. The scaphoid and radial part of the lunate articulates with the radius, and the triquetral and ulnar part of the lunate articulates with the articular disc. A concave surface is formed by the radius and the articular disc.

■ Midcarpal joint
(art. mediocarpalis)

The midcarpal joint is an anatomically simple and mechanically compound joint between the bones of the proximal and distal rows of carpals. The scaphoid has a convex surface distally and articulates with the two trapezii, which together can be considered as having a concave surface. On the ulnar side, the scaphoid, lunate and triquetral form a concave surface which articulates with the convex surface formed by the capitate and hamate.

■ Pisiform joint
(art. ossis pisiformis)

The pisiform joint is an anatomically simple and mechanically compound plane gliding joint. The pisiform is a sesamoid bone in the tendon of the flexor carpi ulnaris. Proximal gliding is prevented by the pisohamate and pisometacarpal ligaments. The abductor digiti minimi arises from the pisiform bone. Therefore, the pisiform will be fixated during contraction of both the abductor digiti minimi and flexor carpi ulnaris.

Wrist

Bony palpation

- Scaphoid, lunate, triquetral, pisiform, radius
- Ulna and articular disc
- Joint spaces between radial-carpal and ulnar-carpal joints
- Intercarpal joints of proximal carpal row
- Trapezium, trapezoid, capitate, hamate with its hook
- Joint spaces between the proximal and distal row of carpals
- Intercarpal joints of distal carpal row

Ligaments

- Intercarpal ligaments: dorsal, palmar and interosseous
- Radiocarpal ligament: deep
- Collateral ligaments: radial and ulnar
- Ulnocarpal ligament: palmar
- Radiocarpal ligaments: dorsal and palmar
- Pisohamate and pisometacarpal ligaments

Bone movement and axes

- **Dorsal flexion** (extension) and **palmar flexion** (flexion): Movement begins in the radiocarpal joint around a transverse axis running through the lunate, and continues in the intercarpal joint around a transverse axis running through the capitate. Approximately half the range of wrist extension and flexion takes place in the intercarpal joint and the other half in the radiocarpal joint.

  During *dorsal flexion,* the proximal part of the capitate moves in a palmar direction in relation to lunate; the same occurs with lunate in relation to the radius. The proximal part of the scaphoid also moves in a palmar direction in relation to the radius; the distal part of the scaphoid appears to move in a palmar direction in relation to the trapezii because these bones glide dorsally on the scaphoid. During *palmar flexion* these movements are reversed.

- **Ulnar flexion** (ulnar deviation, adduction): Movement takes place primarily in the radiocarpal joint around a dorsal-palmar axis through the head of the capitate. The proximal row of carpals glide in a radial direction in relation to the radius. Laxity in the ligaments on the radial side of the joint allows this gliding to take place.
- **Radial flexion** (radial deviation, abduction): The primary movement also takes place around the above mentioned axis and the proximal carpal row glides in an ulnar direction in relation to the radius. However, due to tightening of ligaments, radial flexion (ulnar gliding) is less than ulnar flexion (radial gliding). Full radial flexion requires that the two trapezii glide onto the dorsal side of the scaphoid. This approximates the trapezii and the radius; the movement is similar to that occurring with extension of the wrist.

End feel

- Firm

Joint movement (gliding)

- Convex Rule for all wrist joints with the following exception:
- Concave Rule for trapezium/trapezoid-scaphoid joint

Treatment plane

- On the concave surface of the targeted wrist joint

Zero position

- The longitudinal axes through the radius and the third metacarpal bone form a straight line.

Resting position

- Slight palmar flexion and slight ulnar flexion (midway between maximal radial and ulnar flexion)

Close-packed position

- Wrist in maximal extension

Capsular pattern

- Restricted equally in all directions

■ Wrist examination scheme

(Refer to Chapters 3 and 4 for more information on examination)

Tests of function

1. **Active and passive movements, including stability tests and end-feel**

Palmar flexion	90°
Extension	80°
Radial flexion	20°
Ulnar flexion	30°

2. **Translatoric joint play movements, including end-feel**

Traction - compression	(Figure 8a)
Gliding	
Palmar	(Figure 9a)
Dorsal	(Figure 10a)
Radial	(Figure 11a)
Ulnar	(Figure 12a)
Carpal bones	(Figures 14a)

3. **Resisted movements**

Palmar flexion	OTHER FUNCTIONS
Flexor carpi radialis	Radial flexion
Flexor carpi ulnaris	Ulnar flexion
Palmaris longus	
Extension	
Extensor carpi radialis longus	Radial flexion
Extensor carpi radialis brevis	
Extensor carpi ulnaris	Ulnar flexion
Radial flexion	
Flexor carpi radialis	Palmar flexion
Extensor carpi radialis	Extension
Ulnar flexion	
Flexor carpi ulnaris	Palmar flexion
Extensor carpi ulnaris	Extension

4. **Passive soft tissue movements**

   Physiological

   Accessory

5. **Additional tests**

Trial treatment

Traction	(Figure 8b)

Wrist

Wrist techniques

General techniques

Specific techniques

Capitate-lunate

Lunate-radius

Scaphoid-radius

Trapezii-scaphoid

Triquetral-ulna

Wrist traction
for pain and hypomobility

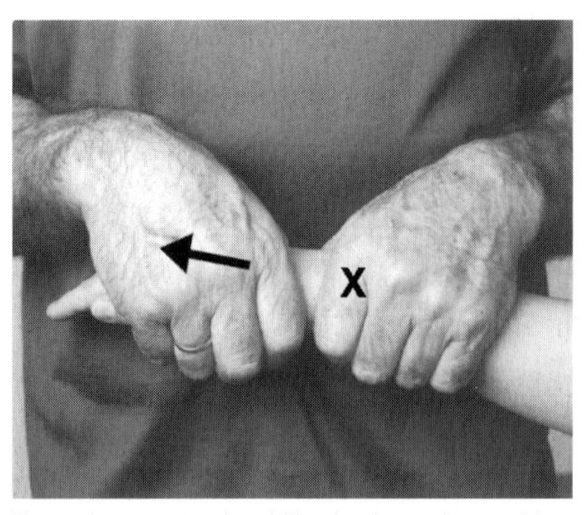

Figure 8a – test and mobilization in resting position

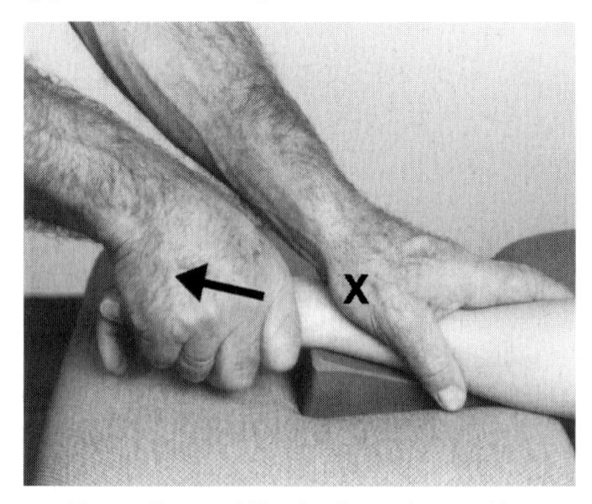

Figure 8b – mobilization in resting position

Wrist

■ Figure 8a: Test and mobilization in resting position

Objective

- To evaluate the quantity and quality of traction joint play in the wrist, including end-feel.
- To decrease pain or increase range-of-motion in the wrist joints.

Starting position

- The patient's palm faces down.
- Position the joint in its resting position.

Hand placement and fixation

- **Therapist's stable hand (left):** Grip the patient's forearm just proximal to the wrist joint; fixate the patient's forearm against your body.
- **Therapist's moving hand (right):** Grip the patient's hand just distal to the wrist joint.

Procedure

- Apply a Grade I, II or III distal traction movement to the distal joint partners.
- Modify your grip to apply more specific traction between the radius and proximal row of carpals, or between the proximal and distal row of carpals.

■ Figure 8b: Mobilization in resting position

- Apply a Grade III traction movement to the wrist joints with the patient's forearm resting on a wedge; fixate the patient's anterior distal forearm against the wedge with your hand; grip with your thenar eminence just proximal to the targeted wrist joints.

Wrist traction

for restricted palmar and dorsal flexion

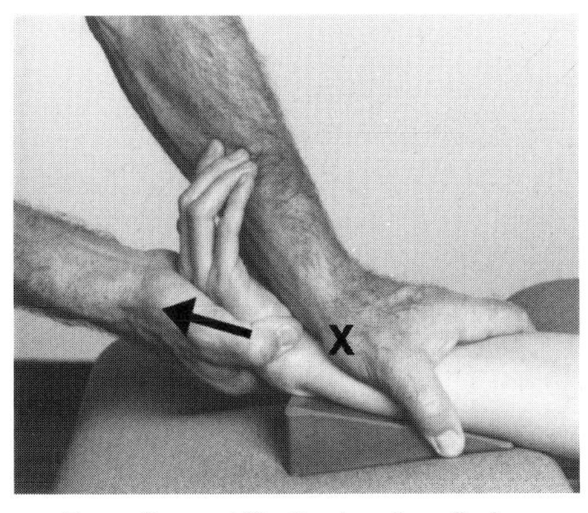

Figure 8c – mobilization in palmar flexion

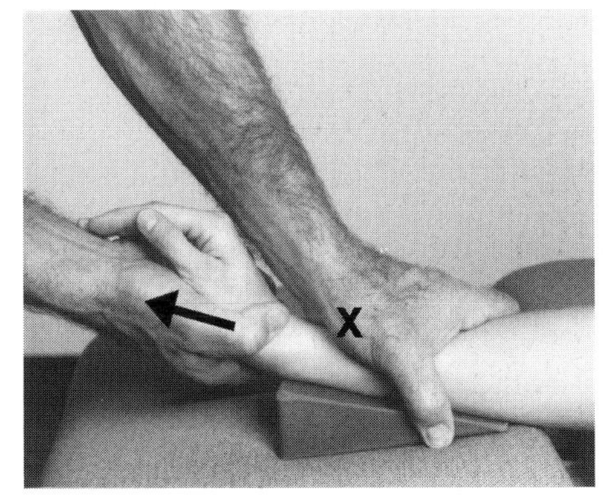

Figure 8d – mobilization in dorsal flexion

■ Figure 8c: Palmar progression

Objective

- To increase wrist palmar flexion range-of-motion.

Starting position

- The posterior side of the patient's distal forearm rests on the wedge.
- Position the joint near the end range-of-motion into palmar flexion.

Hand placement and fixation

- **Therapist's stable hand (left):** Fixate the patient's distal forearm against the wedge, including, if necessary, the proximal row of carpals; grip the patient's distal forearm with your thenar eminence just proximal to the targeted wrist joints.
- **Therapist's moving hand (right):** Grip the patient's hand just distal to the targeted wrist joints.

Procedure

- Apply a Grade III distal traction movement to the distal joint partners.
- Modify your grip to apply more specific traction between the radius and proximal row of carpals, or between the proximal and distal row of carpals.

■ Figure 8d: Dorsal progression

- Apply a Grade III distal traction movement to the wrist positioned close to its end range-of-motion into dorsal flexion.
- The anterior side of the patient's distal forearm rests on the wedge.

Wrist palmar glide
for restricted dorsal flexion

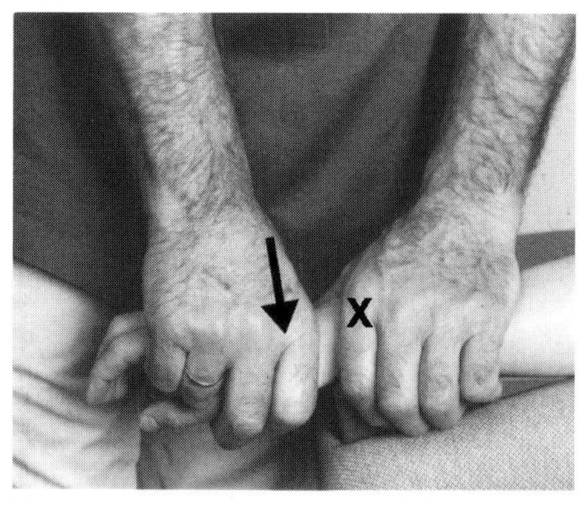

Figure 9a – test and mobilization in resting position

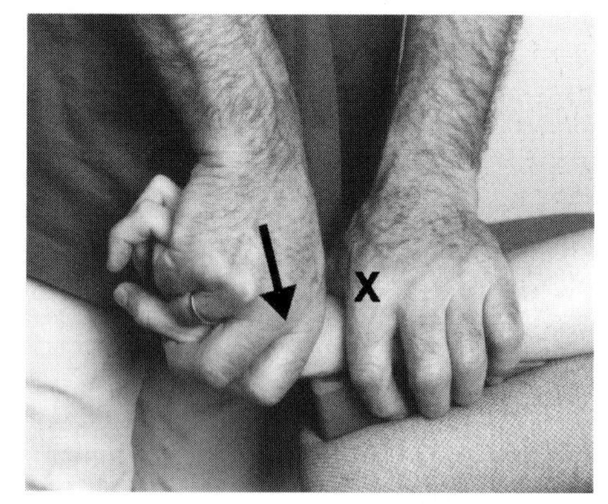

Figure 9b – mobilization in dorsal flexion

Wrist

■ Figure 9a: Test and mobilization in resting position

Objective

- To evaluate the quantity and quality of palmar glide joint play in the wrist joints.
- To increase wrist dorsal flexion range-of-motion (Convex Rule).

Starting position

- The anterior side of the patient's distal forearm rests on the wedge.
- Position the joint in its resting position.

Hand placement and fixation

- **Therapist's stable hand (left):** Fixate the patient's distal forearm against the wedge, including, if necessary, the proximal row of carpals; grip the patient's distal forearm just proximal to the targeted wrist joints.
- **Therapist's moving hand (right):** Hold the patient's hand in your hand; grip just distal to the targeted joint spaces.

Procedure

- Apply a Grade II or III palmar glide movement to the distal joint partners.
- Modify your grip to apply more specific mobilization between the radius and proximal row of carpals, or between the proximal and distal row of carpals.

■ Figure 9b: Dorsal progression

- Apply a Grade III palmar glide movement with the targeted wrist joints positioned close to their end range-of-motion into dorsal flexion (Convex Rule).

Wrist dorsal glide
for restricted palmar flexion

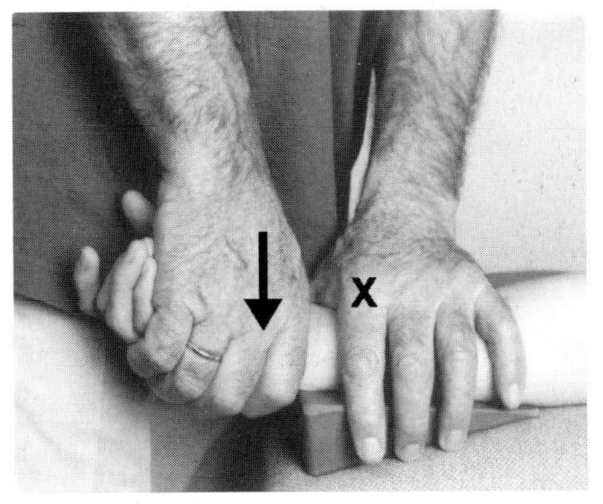

Figure 10a – test and mobilization in resting position

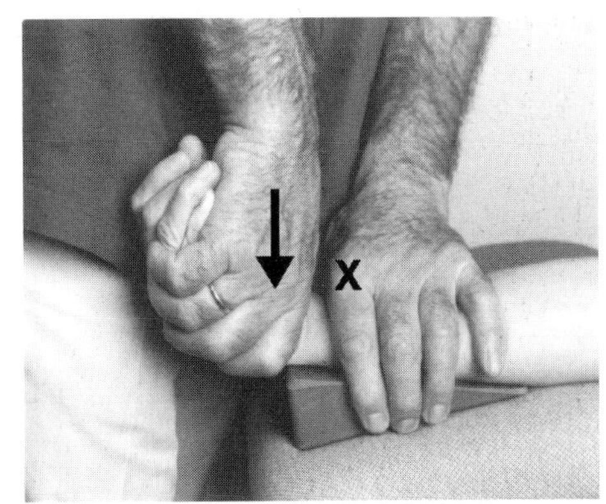

Figure 10b – mobilization in palmar flexion

■ Figure 10a: Test and mobilization in resting position

Objective

- To evaluate the quantity and quality of dorsal glide joint play in the wrist joints.
- To increase wrist palmar flexion range-of-motion (Convex Rule).

Starting position

- The posterior side of the patient's distal forearm rests on the wedge.
- Position the joint in its resting position.

Hand placement and fixation

- **Therapist's stable hand (left):** Fixate the patient's distal forearm against the wedge, including, if necessary, the proximal row of carpals; grip the patient's distal forearm just proximal to the targeted wrist joints.
- **Therapist's moving hand (right):** Hold the patient's hand in your hand; grip just distal to the targeted joints.

Procedure

- Apply a Grade II or III dorsal glide movement to the distal joint partners.
- Modify your grip to apply more specific mobilization between the radius and proximal row of carpals, or between the proximal and distal row of carpals.

■ Figure 10b: Palmar progression

- Apply a Grade III dorsal glide movement with the targeted wrist joints positioned close to their end range-of-motion into palmar flexion (Convex Rule).

Wrist radial glide
for restricted ulnar flexion

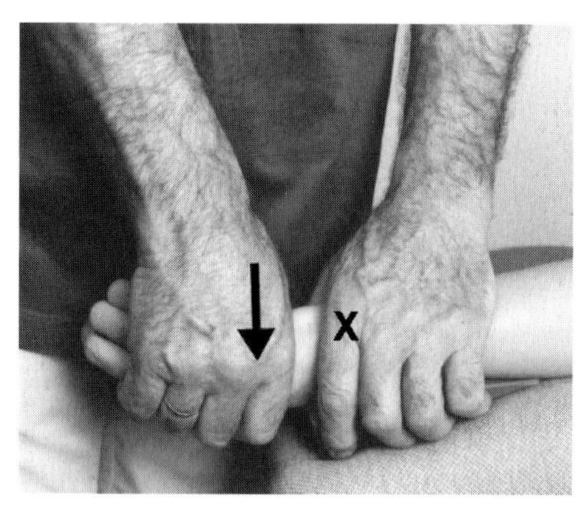

Figure 11a – test and mobilization in resting position

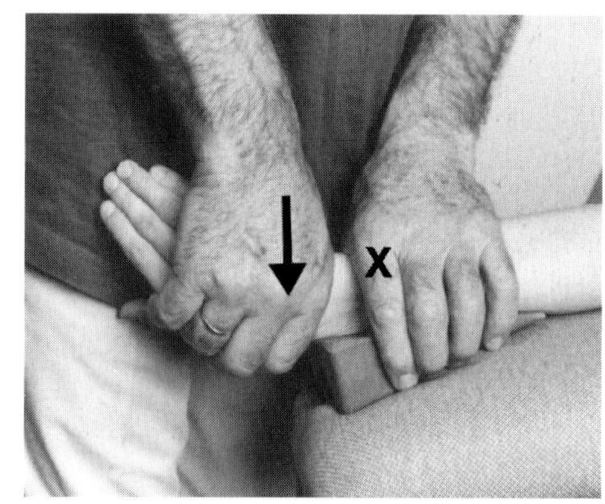

Figure 11b – mobilization in ulnar flexion

■ Figure 11a: Test and mobilization in resting position

Objective

- To evaluate the quantity and quality of radial glide joint play in the wrist joints.
- To increase ulnar flexion range-of-motion (Convex Rule).

Starting position

- The radial side of the patient's distal forearm rests on the wedge.
- Position the joint in its resting position.

Hand placement and fixation

- **Therapist's stable hand (left):** Fixate the patient's distal forearm against the wedge, including, if necessary, the proximal row of carpals; grip the patient's distal forearm just proximal to the targeted wrist joints.
- **Therapist's moving hand (right):** Hold the patient's hand in your hand; grip just distal to the targeted joint spaces.

Procedure

- Apply a Grade II or III radial glide movement to the distal joint partners.

■ Figure 11b: Ulnar progression

- Apply a Grade III radial glide movement with the targeted wrist joints positioned close to their end range-of-motion into ulnar flexion (Convex Rule).

Wrist ulnar glide
for restricted radial flexion

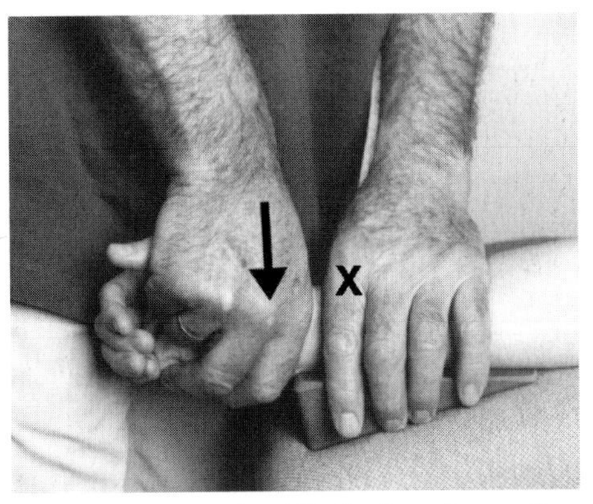

Figure 12a – test and mobilization in resting position

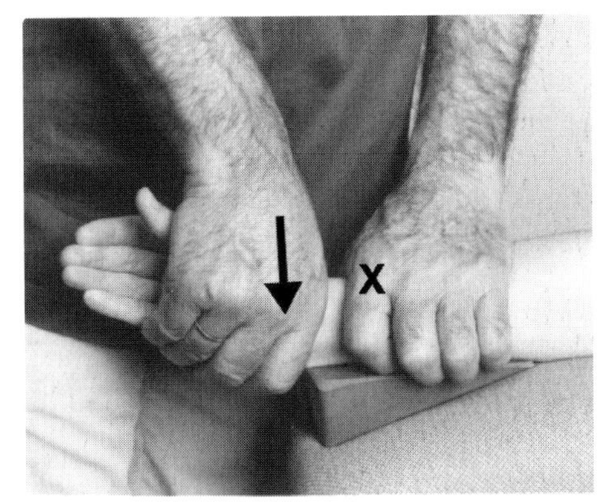

Figure 12b – mobilization in radial flexion

■ Figure 12a: Test and mobilization in resting position

Objective

- To evaluate the quantity and quality of ulnar glide joint play in the wrist joints.
- To increase radial flexion range-of-motion (Convex Rule).

Starting position

- The ulnar side of the patient's distal forearm rests on the wedge.
- Position the joint in its resting position.

Hand placement and fixation

- **Therapist's stable hand (left):** Fixate the patient's distal forearm against the wedge, including, if necessary, the proximal row of carpals; grip the patient's distal forearm just proximal to the targeted wrist joints.
- **Therapist's moving hand (right):** Hold the patient's hand in your hand; grip just distal to the targeted joint spaces.

Procedure

- Apply a Grade II or III ulnar glide movement to the distal joint partners.

■ Figure 12b: Radial progression

- Apply a Grade III ulnar glide movement with the targeted wrist joints positioned close to their end range-of-motion into radial flexion (Convex Rule).

Wrist glide tests
Recommended sequence

■ Figure 13: Recommended glide test sequence for the wrist

Use one hand for fixation and the other hand for movement.

Movements around the capitate (Figure 14a)

Fixate capitate and move:

1. Trapezoid
2. Scaphoid
3. Lunate

Fixate capitate and move:

4. Hamate (adapt from Figure 14a, 15a or 16a)

Movements on the radial side of the wrist

Fixate scaphoid and move:

5. The two trapezii (Figure 15c)

Movements of the radiocarpal joint

Fixate radius and move:

6. Scaphoid (Figure 15a)
7. Lunate (adapt from Figure 14a, 15a or 16a)

Fixate ulna, including the disc, and move:

8. Triquetral (Figure 16a)

Movements on the ulnar side of the wrist

Fixate triquetral and move:

9. Hamate (adapt from Figure 14a, 15a or 16a)
10. Pisiform (position the patient's hand in palmar flexion)

Movements between triquetral-lunate, lunate-scaphoid and between the two trapezii can also be tested.

Wrist

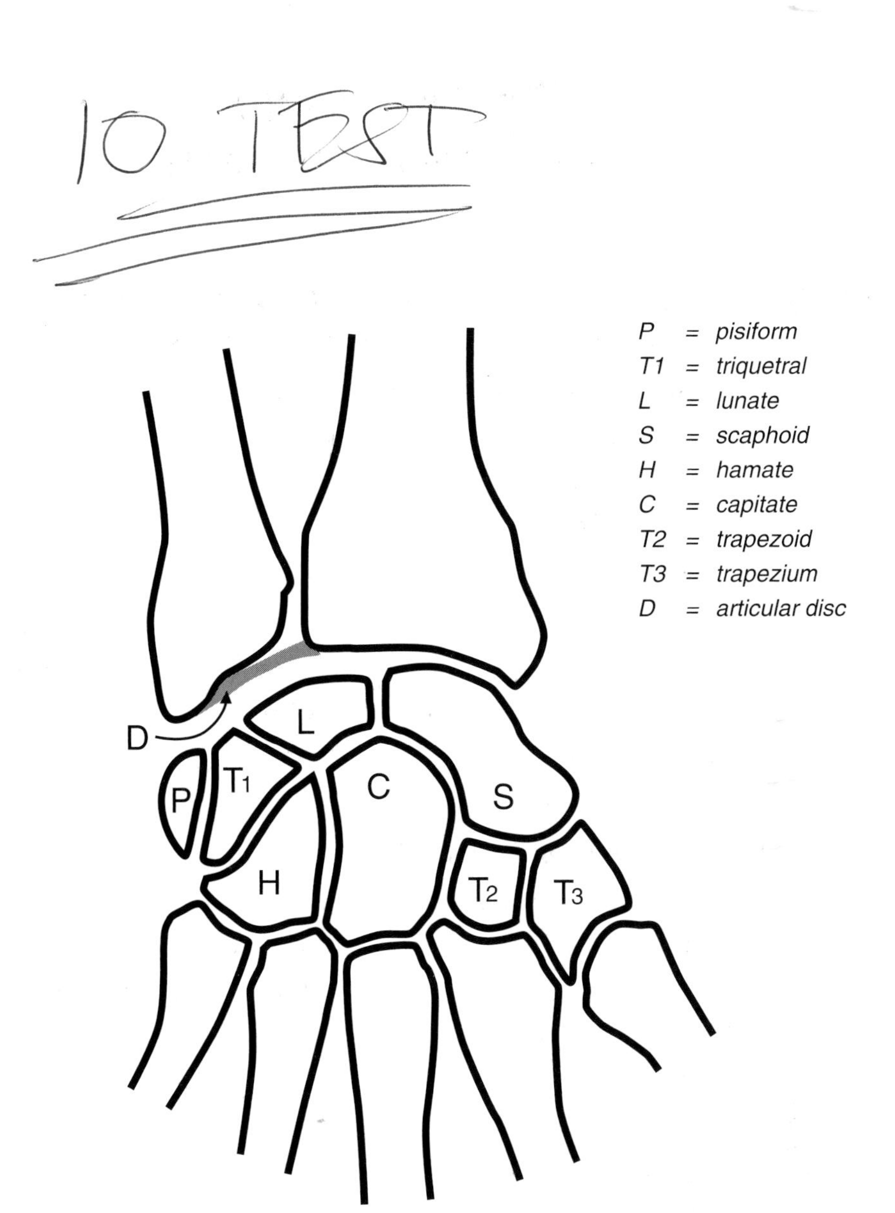

Figure 13
Dorsal aspect of the right wrist

Wrist palmar and dorsal glide

Test

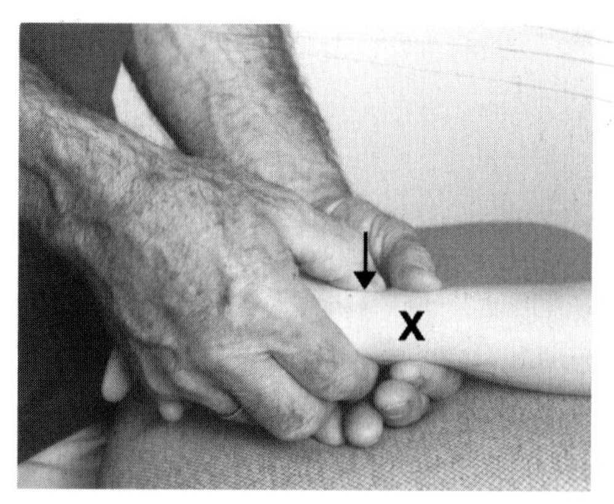

Figure 14a - specific wrist joint tests

■ Figure 14a: Test with proximal fixation in resting position

Objective

- To evaluate the quantity and quality of palmar and dorsal glide joint play in specific wrist joints, including end-feel.

Starting position

- The patient rests the palmar side of their forearm on the treatment surface.
- Position the joint in its resting position.

Fixation

- **Therapist's stable hand (left):** Grip the proximal joint partner with your fingers; fixate your hand against the treatment surface.
- **Therapist's moving hand (right):** Hold the patient's fingers in your hand; grip with your fingers just distal to the targeted joint space.

Procedure

- Apply a palmar or dorsal glide movement; use simultaneous Grade I traction to facilitate the movement.

■ Distal or lateral fixation in resting position (not shown)

- All joints between the eight carpal bones can be similarly tested using proximal, distal, or lateral fixation: fixate one carpal bone and move an adjacent carpal bone in a dorsal or palmar direction.

Capitate-lunate palmar glide
for restricted dorsal flexion

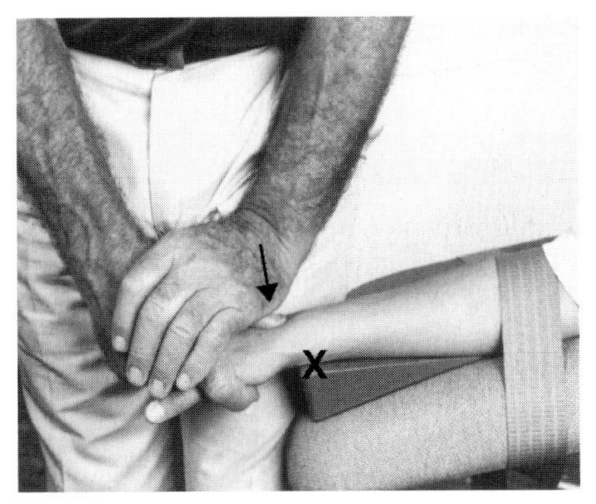

Figure 14b – mobilization in resting position

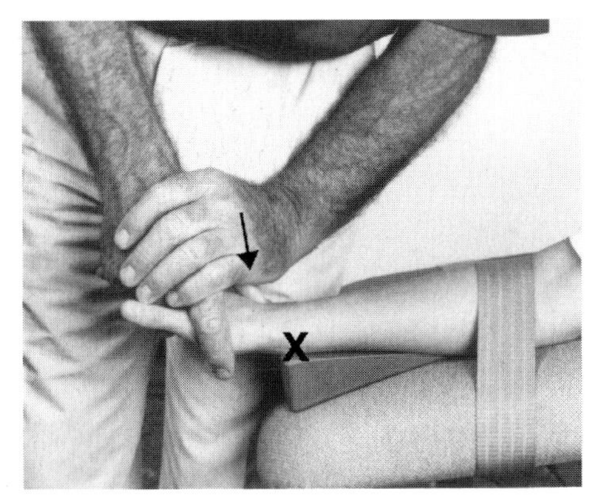

Figure 14c – mobilization in dorsal flexion

■ Figure 14b: Mobilization in resting position

Objective

- To increase wrist dorsal flexion range-of-motion (Convex Rule).

Starting position

- The anterior side of the patient's distal forearm, including the lunate, rests on the wedge.
- Position the joint in its resting position.

Hand placement and fixation

- **Fixation:** The lunate is fixated by the wedge; a belt fixates the proximal forearm.
- **Therapist's moving hands:** Support the patient's hand and thumb in your right hand with your thumb on the capitate; use your left hand to supplement your grip.

Procedure

- Apply a Grade III palmar glide movement to the capitate.

■ Figure 14c: Dorsal flexion progression

- Apply a Grade III palmar glide joint play movement to the capitate near the end range-of-motion into wrist dorsal flexion (Convex Rule)

■ Lunate-radius mobilization (not shown)

- Fixate the radius; apply a Grade III palmar glide movement to the lunate (Convex Rule).

Capitate-lunate dorsal glide
for restricted palmar flexion

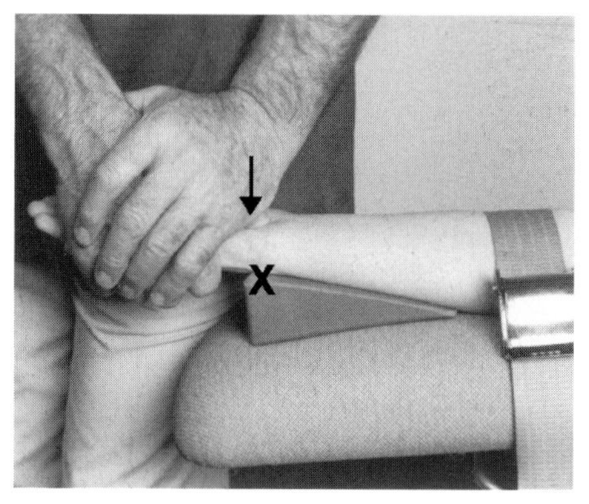

Figure 14d – mobilization in resting position

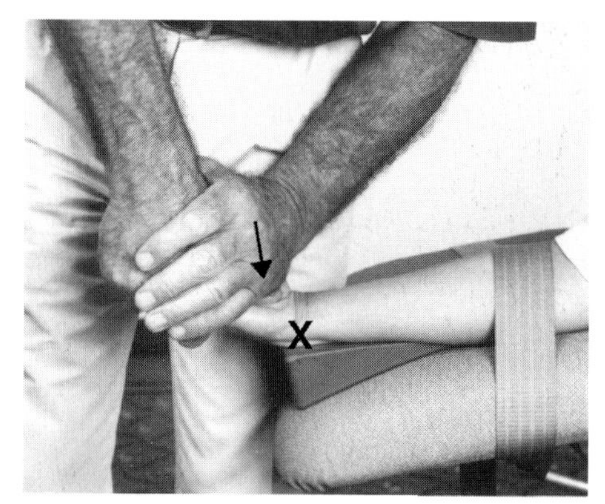

Figure 14e – mobilization in palmar flexion

■ Figure 14d: Mobilization in resting position

Objective

- To increase wrist palmar flexion range-of-motion (Convex Rule).

Starting position

- The posterior side of the patient's distal forearm, including the lunate, rests on the wedge.
- Position the joint in its resting position.

Hand placement and fixation

- **Fixation:** The lunate is fixated by the wedge; a belt fixates the proximal forearm.
- **Therapist's moving hands:** Support the patient's hand and thumb in your right hand with your thumb on the capitate; use your left hand to supplement your grip.

Procedure

- Apply a Grade III dorsal glide movement to the capitate.

■ Figure 14e: Palmar flexion progression

- Apply a Grade III dorsal glide joint play movement near the end range-of-motion into wrist palmar flexion (Convex Rule).

■ Radius-lunate mobilization (not shown)

- Fixate the radius; apply a Grade III dorsal glide movement to the lunate (Convex Rule).

Scaphoid-radius palmar glide
for restricted dorsal flexion

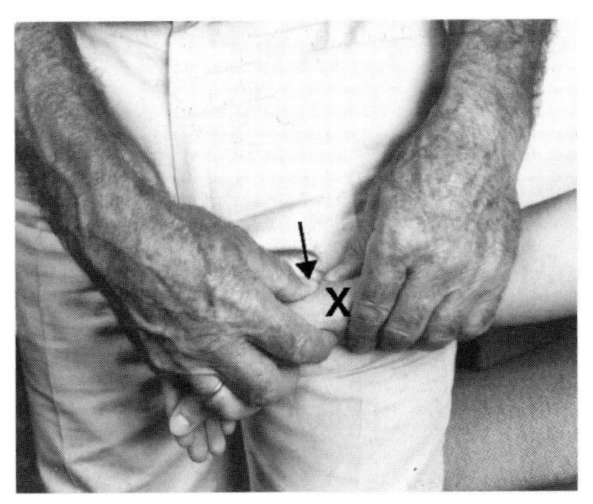

Figure 15a – test and mobilization in resting position

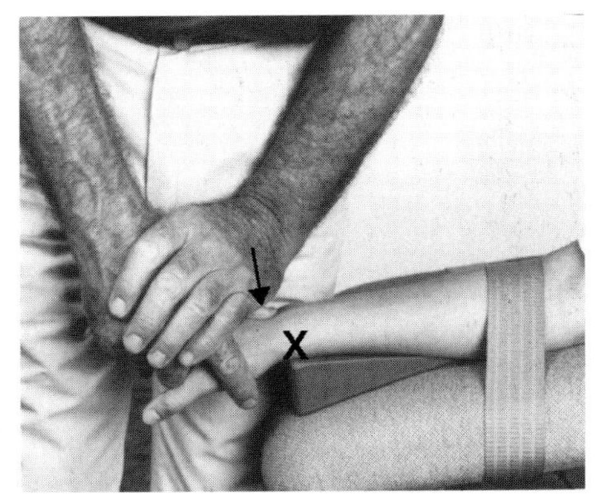

Figure 15b – mobilization in resting position

Wrist

■ Figure 15a: Test and mobilization in resting position

Objective

- To evaluate the quantity and quality of scaphoid palmar glide joint play in relation to the radius, including end-feel.
- To increase wrist dorsal and radial flexion range-of-motion (Convex Rule).

Starting position

- The anterior side of the patient's forearm faces down.
- Position the joint in its resting position.

Hand placement and fixation

- **Therapist's stable hand (left):** Hold the patient's distal forearm against your body; grip with your fingers just proximal to the scaphoid-radius joint space.
- **Therapist's moving hand (right):** Support the patient's hand in your hand; grip with your thumb and index finger surrounding the scaphoid.

Procedure

- Apply a Grade II or III palmar glide movement to the scaphoid.

■ Figure 15b: Mobilization in resting position

- The anterior side of the patient's distal forearm rests on the wedge; the distal radius is fixated by the wedge; a belt fixates the proximal forearm; support the patient's hand and thumb in your right hand with your thumb on the scaphoid; use your left hand to supplement your grip; apply a Grade III palmar glide movement to the scaphoid near the end range-of-motion into wrist dorsal flexion.

■ Trapezii-scaphoid palmar glide for restricted palmar flexion (not shown)

- Fixate the scaphoid; apply a Grade III palmar glide movement to the trapezii (Concave Rule).

Trapezii-scaphoid dorsal glide
for restricted dorsal flexion

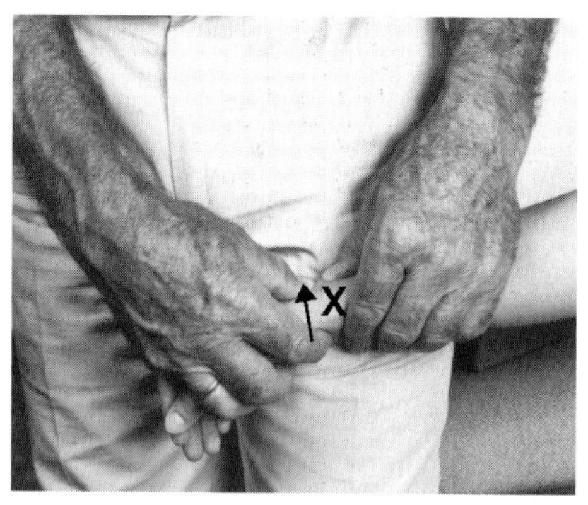

Figure 15c – test and mobilization in resting position

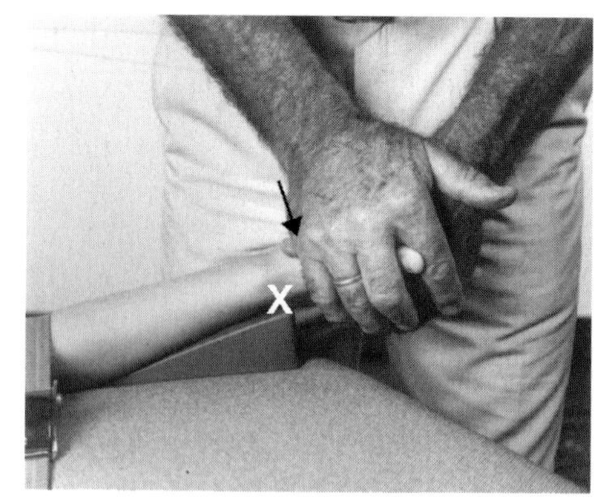

Figure 15d – mobilization in resting position

Wrist

■ Figure 15c: Test and mobilization in resting position

Objective

- To evaluate the quantity and quality of dorsal glide joint play of the trapezii in relation to the scaphoid, including end-feel.
- To increase wrist dorsal flexion (Concave Rule).

Starting position

- The anterior side of the patient's forearm faces down.
- Position the joint in its resting position.

Hand placement and fixation

- **Therapist's stable hand (left):** Hold the patient's distal forearm against your body; grip with your fingers surrounding the scaphoid.
- **Therapist's moving hand (right):** Support the patient's hand in your hand; grip with your thumb and index finger surrounding the trapezii.

Procedure

- Apply a Grade II or III dorsal glide movement to the trapezii.

■ Figure 15d: Mobilization in resting position

- The posterior side of the patient's distal forearm, including the scaphoid, rests on the wedge; the distal radius is fixated by the wedge; a belt fixates the proximal forearm; support the patient's hand and thumb in your left hand with your thumb on the trapezii; use your right hand to supplement your grip; apply a Grade III dorsal glide movement to the trapezii.

■ Scaphoid-radius dorsal glide for restricted dorsal flexion (not shown)

- Fixate the radius; apply a Grade III dorsal glide movement to the scaphoid (Convex Rule).

Triquetral-ulna palmar glide

for restricted dorsal flexion

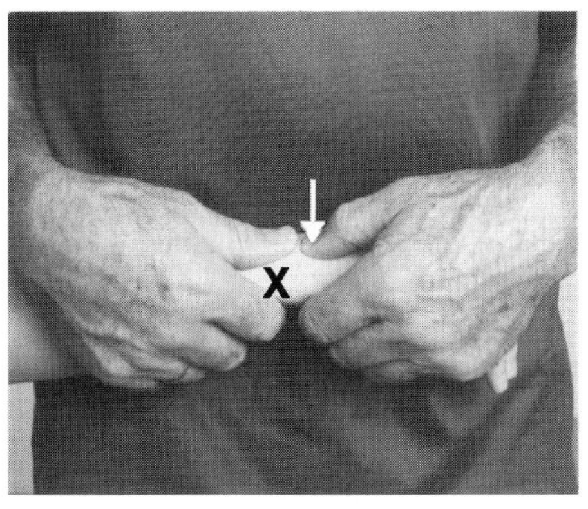

Figure 16a – test and mobilization in resting position

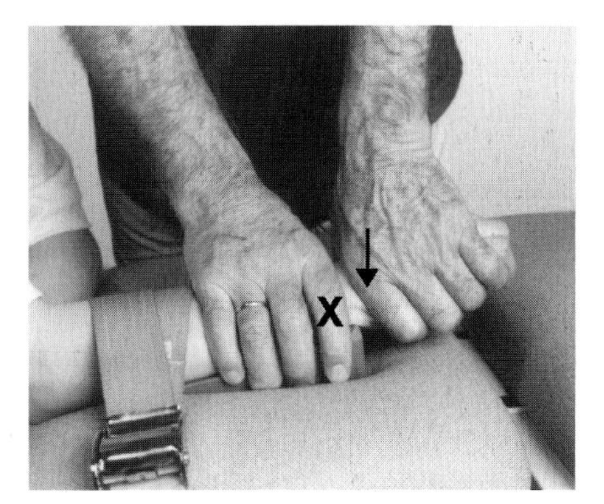

Figure 16b – mobilization in resting position

■ Figure 16a: Test and mobilization in resting position

Objective

- To evaluate the quantity and quality of palmar glide joint play of the triquetral in relation to the ulna, including end-feel.
- To release a fixated articular disc between triquetral and ulna. The fixated articular disc can restrict forearm pronation, supination, and all wrist movements.
- To increase wrist dorsal flexion (Convex Rule).

Starting position

- The palmar side of the patient's hand faces down.
- Position the joint in its resting position.

Hand placement and fixation

- **Therapist's stable hand (right):** Hold the patient's distal forearm against your body; grip with your fingers around the head of the ulna.
- **Therapist's moving hand (left):** Support the patient's hand in your hand; grip with your thumb and index finger surrounding the triquetral.

Procedure

- Apply a Grade II or III palmar glide movement to the triquetral.

■ Figure 16b: Mobilization in resting position

- The anterior side of the patient's distal forearm rests on the wedge; the distal radius is fixated by the wedge; a belt fixates the proximal forearm; support the patient's hand in your right hand with your MCP II joint on the triquetral; apply a Grade III palmar glide joint play movement of the triquetral.

■ Hamate-triquetral palmar glide (not shown)

- Fixate the triquetral; apply a Grade III palmar glide movement to the hamate.

■ Hamate-triquetral dorsal glide (not shown)

- Supinate the patient's forearm and fixate the triquetral; apply a Grade III dorsal glide movement to the hamate.

Notes

Wrist

CHAPTER 11

FOREARM

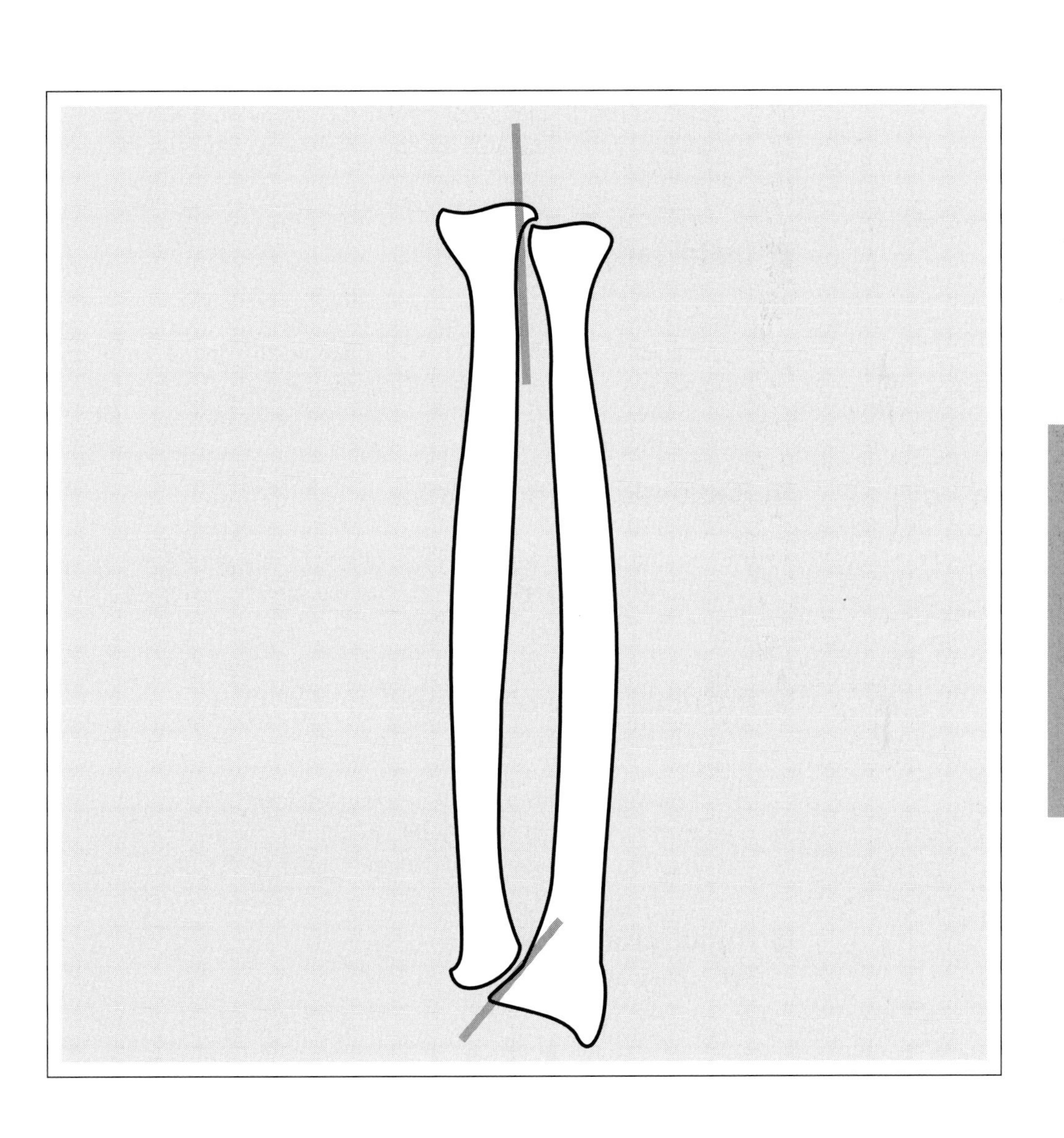

11 Forearm

Functional anatomy and movement

The forearm (antebrachium) consists of the radius and ulna with the antebrachial interosseous membrane. The distal broad part of the radius has the main contact with the carpus while the proximal, thickened part of the ulna has the main connection with the upper arm.

Distal radio-ulnar joint

(art. radio-ulnaris distalis)

The distal radio-ulnar joint is an anatomically and mechanically simple biaxial joint (trochoid, modified ovoid). The head of the ulna moves in the notch on the distal end of the radius.

Radio-ulnar syndesmosis

The radio-ulnar syndesmosis involves the length of the ulna and radius with the antebrachial interosseous membrane lying between their sharp interosseous borders.

Proximal radio-ulnar joint

(art. radio-ulnaris proximalis)

The proximal radio-ulnar joint is anatomically part of the elbow joint; it is a biaxial pivot joint (trochoid, modified ovoid). The head of the radius moves in the radial notch of the ulna.

Humeroradial joint

(art. humeroradialis)

The humeroradial joint is anatomically part of the elbow, but functionally also part of the forearm, and so is described here.

The humeroradial joint is a triaxial joint (spheroid, unmodified ovoid). During flexion-extension, the shallow concave facet on the radius moves on the convex surface of the capitulum of the humerus. Movements also take place at this joint during pronation and supination of the forearm. Testing and treatment of the humeroradial joint is described in *Chapter 12: Elbow.*

Bony palpation

- Proximal radius (radial head)
- Radial tuberosity
- Distal radius and styloid process
- Distal ulna and styloid process
- Distal radio-ulnar joint space
- Proximal radio-ulnar joint space
- Humeroradial joint space

Ligaments

- Annular ligament: The annular ligament is cone-shaped, narrows distally and is only attached to the ulna which allows free movement of the head of the radius.
- Radial collateral ligament

Bone movement and axes

- Pronation - supination: The radius rotates around the ulna and produces torsion of the forearm; the axis of movement lies obliquely in the forearm passing through the radial and ulnar heads.
- Abduction (passive): The radius glides distally in relation to the ulna.
- Adduction (passive): The radius glides proximally in relation to the ulna.

End feel

- Pronation: hard end-feel. Pronation stops when the radius comes in contact with the ulna (bone to bone) producing a hard end-feel.
- Supination: firm end-feel. Supination is limited by soft tissues being stretched, especially ligaments, which results in a firm end-feel.

Joint movement (gliding)

- Distal radio-ulnar joint: Concave Rule
- Proximal radio-ulnar joint: Convex Rule
- Radius-humerus: Concave Rule

Treatment plane

- Distal radio-ulnar joint: on the concave joint surface of the radius
- Proximal radio-ulnar joint: on the concave joint surface of the ulna
- Humeroradial joint: on the concave joint surface of the radius

Zero position

- Distal and proximal radio-ulnar joints: upper arm parallel to the trunk with the elbow at a right angle, wrist in the Zero Position, and hand in the sagittal plane
- Humeroradial joint: arm and forearm in the frontal plane with the forearm fully supinated and the elbow extended

Resting position

All joints in the forearm cannot be placed in the resting position simultaneously.

- Distal radio-ulnar joint: The forearm is supinated approximately 10°.
- Proximal radio-ulnar joint: Forearm supination is approximately 35° and elbow flexion approximately 70°.
- Humeroradial joint: The forearm is fully supinated and the elbow fully extended.

Close-packed position

- Distal and proximal radio-ulnar joints: maximal pronation or supination
- Humeroradial joint: 90° elbow flexion

Capsular pattern

- Pronation and supination are restricted equally; occurs usually only when there is marked limitation of flexion and extension of the elbow joint.

■ Forearm examination scheme

(Refer to Chapters 3 and 4 for more information on examination)

Tests of function

A. Active and passive movements, including stability tests and end-feel

Pronation	80°
Supination	90°

B. Translatoric joint play movements

Traction - compression	(adapt techniques from Figure 17)
Gliding	
Distal radio-ulnar joint	
Ventral	(Figure 17a)
Dorsal	(Figure 18a)
Proximal radio-ulnar joint	
Ventral	(Figure 19a)
Dorsal	(Figure 19d)
Humeroradial joint	
Bilateral	(Figure 20)
Dorsal	(Figure 21a)

C. Resisted movements

	OTHER FUNCTIONS
Pronation	
Pronator teres	Flexion
Pronator quadratus	
Brachioradialis	Flexion; functions from supinated to zero starting position
Supination	
Supinator	Extension
Biceps brachii	Flexion
Brachioradialis	Flexion; functions from pronated to zero starting position

D. Passive soft tissue movements

Physiological

Accessory

E. Additional tests

Trial treatment

Distal radio-ulnar joint	
Anterior-posterior gliding	(Figure 17b,18b)
Proximal radio-ulnar joint	
Anterior-posterior gliding	(Figure 19b, 19e)
Radius-humerus	
Posterior gliding	(Figure 21b)

Forearm

Forearm techniques

Distal radio-ulnar joint

Proximal radio-ulnar joint

Humeroradial joint

Radio-ulnar joint

Distal radio-ulnar joint anterior glide
for hypomobility

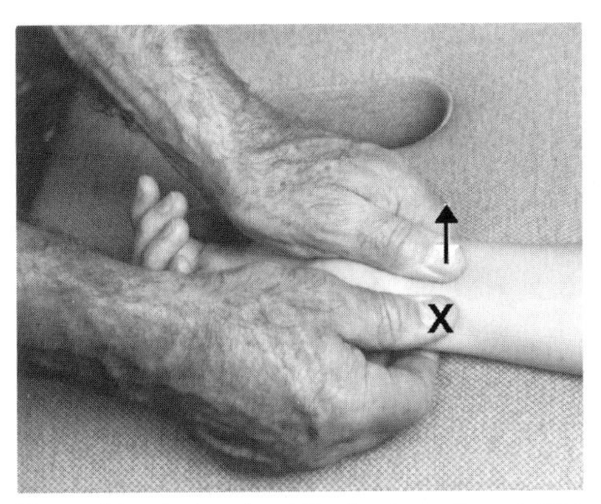

Figure 17a – test and mobilization in resting position

■ Figure 17a: Test and mobilization in resting position

Objective

- To evaluate the quantity and quality of anterior glide joint play in the radio-ulnar joint, including end-feel.
- To increase forearm pronation (Concave Rule).

Starting position

- The posterior side of the patient's forearm rests on the treatment surface, the elbow slightly flexed.
- Position the joint in its resting position.

Hand placement and fixation

- **Therapist's stable hand (right):** Hold the patient's hand from the ulnar side; grip around the patient's distal ulna near the joint space.
- **Therapist's moving hand (left):** Hold the patient's hand from the radial side; grip around the patient's distal radius near the joint space.

Procedure

- Apply a Grade II or III anterior glide movement to the radius.

Note

- Passive pronation, which is greater with the elbow flexed than extended, implicates a shortened supinator.

Distal radio-ulnar joint anterior glide
for restricted pronation

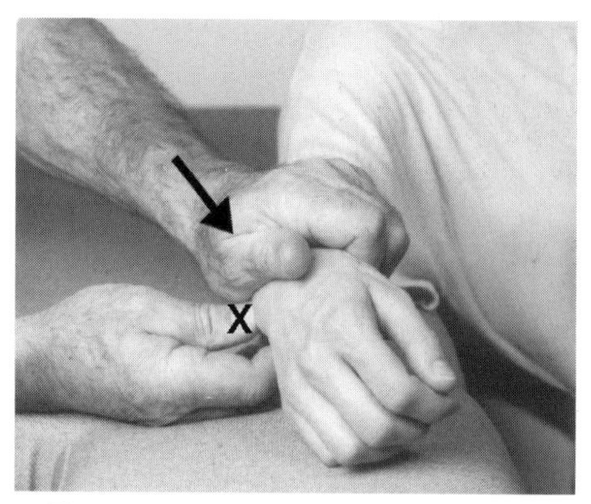

Figure 17b – mobilization in resting position

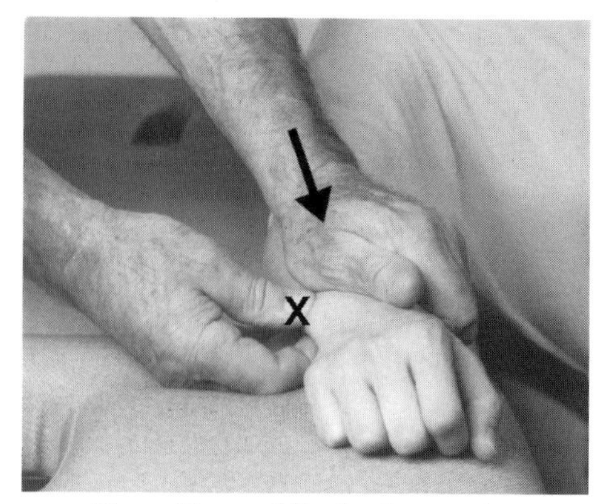

Figure 17c – mobilization in pronation

■ Figure 17b: Mobilization in resting position

Objective

- To increase forearm pronation (Concave Rule).

Starting position

- The ulnar side of the patient's forearm rests on the treatment surface, the elbow slightly flexed.
- Position the joint in its resting position.

Hand placement and fixation

- **Therapist's stable hand (right):** Grip with your thumb and fingers around the patient's ulnar head; rest your forearm on the treatment surface.
- **Therapist's moving hand (left):** Hold the patient's distal forearm from the radial side; grip around the distal radius with your thenar eminence close to the joint space; position your forearm in line with the treatment plane.

Procedure

- Press your left hand downward to apply a Grade III anterior glide movement.

■ Figure 17c: Pronation progression

- Apply a Grade III anterior glide movement with the forearm positioned near its end range-of-motion into pronation.

Distal radio-ulnar joint posterior glide

for hypomobility

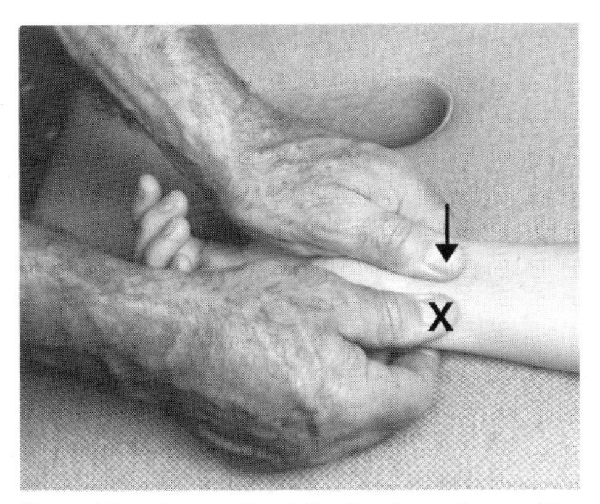

Figure 18a – test and mobilization in resting position

■ Figure 18a: Test and mobilization in resting position

Objective

- To evaluate the quantity and quality of posterior glide movement in the radio-ulnar joint, including end-feel.
- To increase forearm supination (Concave Rule).

Starting position

- The posterior side of the patient's forearm rests on the treatment surface, the elbow slightly flexed.
- Position the joint in its resting position. Note that in positions of greater shoulder abduction, the patient's hand moves into the horizontal plane making it difficult to perform these procedures in the resting position.

Hand placement and fixation

- **Therapist's stable hand (right):** Hold the patient's hand and distal forearm from the ulnar side; grip around the patient's distal ulna near the joint space.
- **Therapist's moving hand (left):** Hold the patient's hand and distal forearm from the radial side; grip around the patient's distal radius near the joint space.

Procedure

- Apply a Grade II or III posterior glide movement to the radius.

Note

- Passive supination, which is greater with the elbow flexed than extended, implicates a shortened pronator teres.

Distal radio-ulnar joint posterior glide

for restricted supination

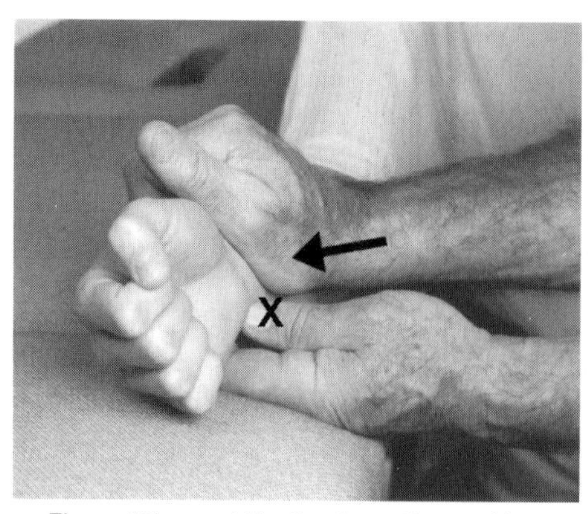

Figure 18b – mobilization in resting position

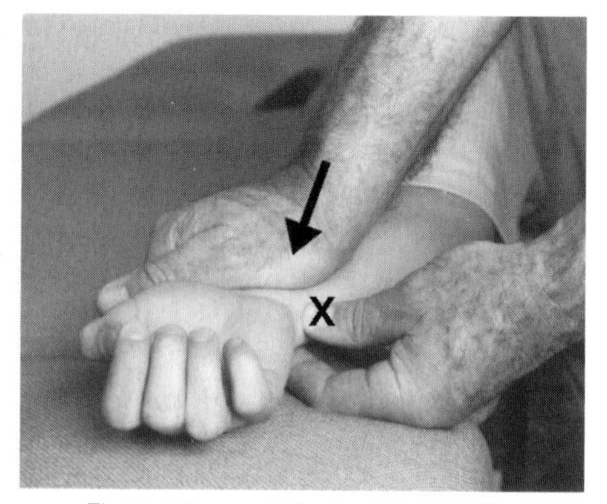

Figure 18c – mobilization in supination

■ Figure 18b: Mobilization in resting position

Objective

- To increase forearm supination (Concave Rule).

Starting position

- The ulnar side of the patient's forearm rests on the treatment surface, the elbow slightly flexed.
- Position the joint in its resting position.

Hand placement and fixation

- **Therapist's stable hand (left):** Grip with your thumb and fingers around the patient's ulnar head; rest your hand on the treatment surface.
- **Therapist's moving hand (right):** Hold the patient's distal forearm from the radial side; grip around the distal radius with your thenar eminence close to the joint space; position your forearm in line with the treatment plane.

Procedure

- Apply a Grade III posterior glide movement to the radius by pressing your right hand in a posterior direction.

■ Figure 18c: Supination progression

- Apply a Grade III posterior glide movement to the radius with the forearm positioned near its end range-of-motion into supination.

Proximal radio-ulnar joint anterior glide
for restricted supination

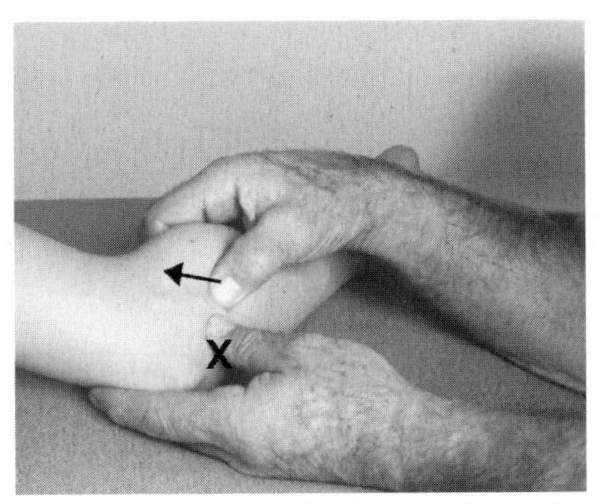

Figure 19a – test and mobilization in resting position

■ Figure 19a: Test and mobilization in resting position

Objective

- To evaluate the quantity and quality of anterior glide joint play in the proximal radio-ulnar joint, including end-feel.
- To increase forearm supination (Convex Rule).

Starting position

- The ulnar side of the patient's forearm rests on the treatment surface, the elbow flexed, shoulder in abduction.
- Position the joint in its resting position.

Hand placement and fixation

- **Therapist's stable hand (left):** Hold the patient's proximal forearm from the ulnar side; grip around the patient's proximal ulna with your palpating thumb in the joint space.
- **Therapist's moving hand (right):** Hold the patient's proximal forearm from the radial side; grip around the patient's proximal radius near the joint space.

Procedure

- Apply a Grade II or III anterior glide movement to the radius.

Proximal radio-ulnar joint anterior glide
for restricted supination (cont'd)

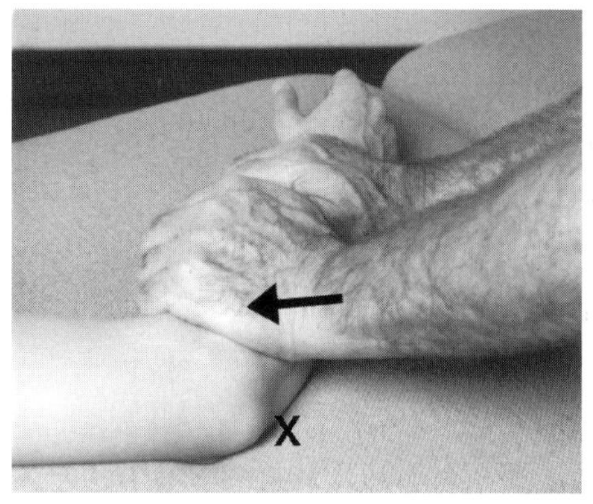

Figure 19b – mobilization in resting position

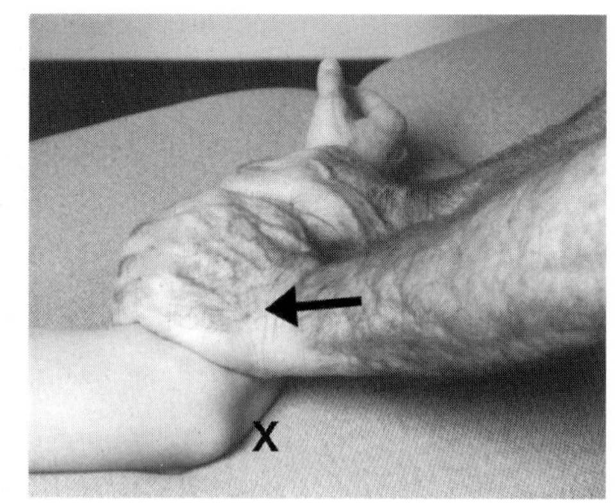

Figure 19c - mobilization in supination

■ Figure 19b: Mobilization in resting position

Objective

- To increase forearm supination (Convex Rule)

Starting position

- The ulnar side of the patient's forearm rests on the treatment surface, the elbow flexed, shoulder in abduction.
- Position the joint in its resting position.

Hand placement and fixation

- **Therapist's stable hand (left):** Hold the patient's forearm from the ulnar or radial side; fixate the patient's forearm against the treatment surface.
- **Therapist's moving hand (right):** Hold the patient's proximal radius with your hand and fingers, with your hypothenar eminence near the joint space; position your forearm in line with the treatment plane.

Procedure

- Apply a Grade III anterior glide movement to the proximal radius. To keep your movement in the treatment plane, you must keep your forearm close to the treatment surface. If you lift your forearm and press downward you will compress the proximal radio-ulnar joint.

■ Figure 19c: Supination progression

- Position the forearm near the end range-of-motion into supination.

Proximal radio-ulnar joint posterior glide
for restricted pronation

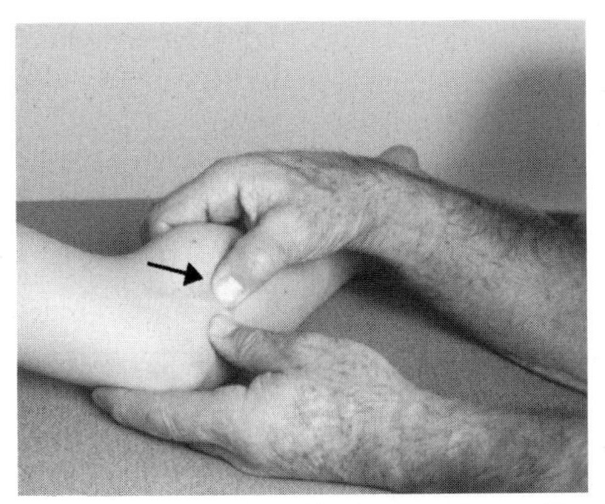

Figure 19d – test and mobilization in resting position

■ Figure 19d: Test and mobilization in resting position

Objective

- To evaluate the quantity and quality of posterior glide joint play in the proximal radio-ulnar joint, including end-feel.
- To increase forearm pronation (Convex Rule).

Starting position

- The ulnar side of the patient's forearm rests on the treatment surface, the elbow flexed, shoulder in abduction.
- Position the joint in its resting position.

Hand placement and fixation

- **Therapist's stable hand (left):** Hold the patient's proximal forearm from the ulnar side; grip around the patient's proximal ulna with your palpating thumb in the joint space.
- **Therapist's moving hand (right):** Hold the patient's proximal forearm from the radial side; grip around the patient's proximal radius near the joint space.

Procedure

- Apply a Grade II or III posterior glide movement to the radius.

Proximal radio-ulnar joint posterior glide

for restricted pronation (cont'd)

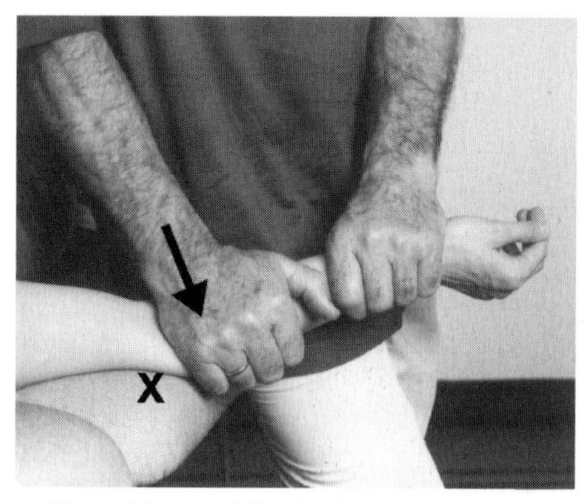

Figure 19e – mobilization in resting position

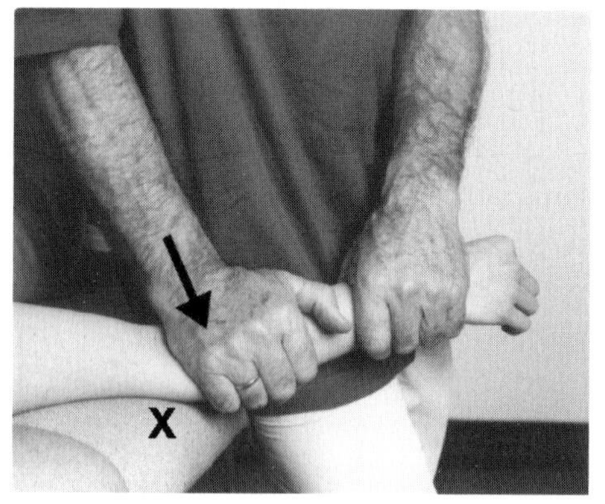

Figure 19f - mobilization in pronation

■ Figure 19e: Mobilization in resting position

Objective

- To increase forearm pronation (Convex Rule)

Starting position

- The posterior side of the patient's arm and proximal ulna rests on the treatment surface, the forearm extends over the edge.
- Position the joint in its resting position.

Hand placement and fixation

- **Fixation:** The ulna is fixated against the treatment surface.
- **Therapist's moving hands:** Hold the patient's radius with both hands; grip with your right hypothenar eminence near the joint space; position your right forearm in line with the treatment plane.

Procedure

- Apply a Grade III posterior glide movement to the proximal radius by bending your knees and leaning your body through your extended arm.

■ Figure 19f: Pronation progression

- Position the forearm near the end range-of-motion into pronation.

Humeroradial joint

test

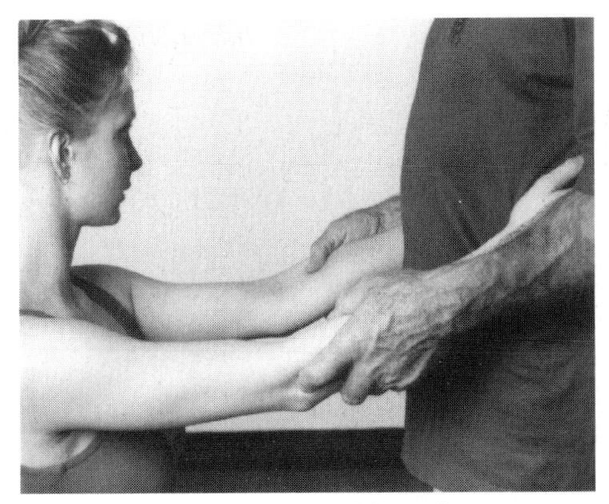

Figure 20 – test bilateral

■ Figure 20: Test bilateral

Objective

- **Position test** for the radial head in relation to the capitulum of the humerus
- **Mobility test** for the humeroradial joint

Starting position

- The patient extends both arms forward.

Hand placement and fixation

- **Fixation**: Hold the patient's forearms against your body with your forearms; the patient remains relaxed, so that the soft tissues crossing the joint do not interfere with the test.
- **Therapist's moving hands**: Grip the patient's proximal forearms from the radial side with your palpating index fingers in the humeroradial joint space.

Procedure

- **Position test**: Palpate the distance between the radial head and capitulum of the humerus; palpate from all sides: posterior, lateral, and ventral.
- **Mobility test**: Flex and extend, abduct and adduct the patient's elbows while you palpate the joint space.

Radius-humerus posterior glide
for restricted extension

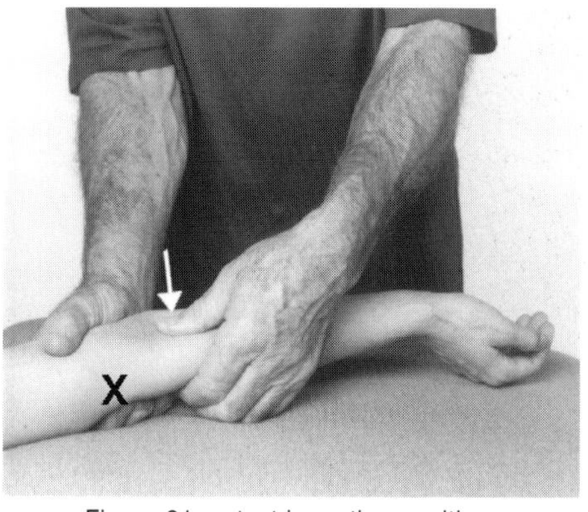

Figure 21a – test in resting position

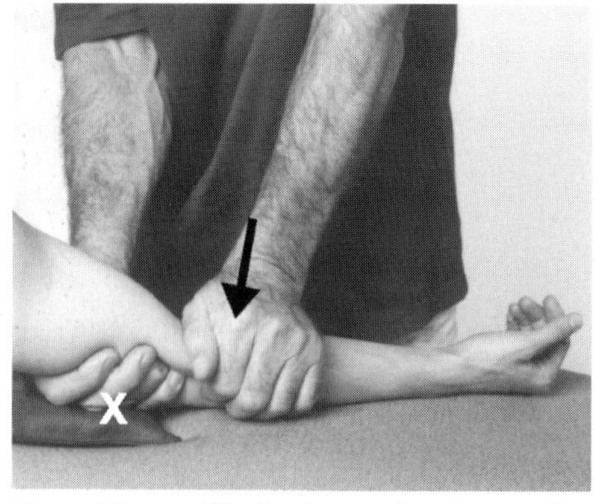

Figure 21b - mobilization in actual resting position

■ Figure 21a: Test in resting position

Objective

- To evaluate the quantity and quality of posterior glide movement in the humeroradial joint, including end-feel. Restricted radius-humerus posterior glide affects elbow extension (Concave Rule).

Starting position

- The posterior side of the patient's arm and forearm rests on the treatment surface, the elbow extended as far as possible.

Hand placement and fixation

- **Therapist's stable hand (right):** Hold the distal arm; fixate the distal humerus against the treatment surface.
- **Therapist's moving hand (left):** Hold the patient's proximal radius; grip around the radial head with your thumb and fingers.

Procedure

- Apply a Grade II or III posterior glide movement to the proximal radius.

■ Figure 21b: Mobilization in actual resting position

- Apply a Grade III posterior glide movement to the radius to increase elbow extension. Position the elbow near the end range-of-motion into elbow extension. Since the resting position of the humeroradial joint is in full extension, with restricted elbow extension the patient will not be able to achieve the resting position.

■ Anterior glide mobilization (not shown)

- Apply a Grade III anterior glide movement to the proximal radius to increase elbow flexion.

Radio-ulnar joint distal glide

for elbow and forearm hypomobility

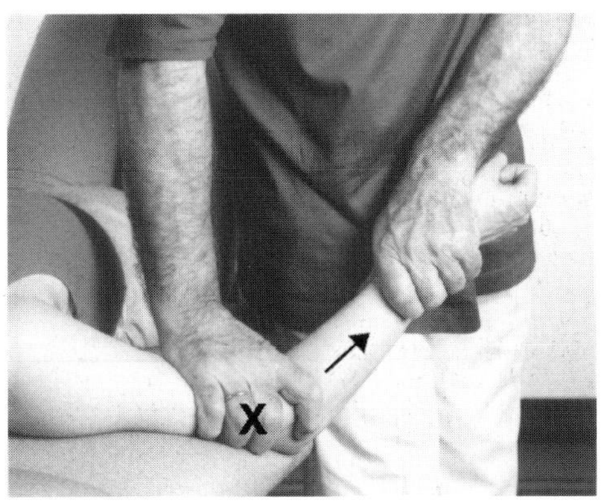

Figure 22 – mobilization in resting position

■ Figure 22: Mobilization in resting position

Objective

- To increase forearm pronation and supination, as well as elbow flexion and extension; distal glide movement of the radius in relation to the ulna stretches the syndesmosis.
- To correct a proximal positional fault of the radius in relation to the humerus by moving the radius distally. See figure 20 for position testing technique.

Starting position

- The posterior side of the patient's arm and proximal ulna rests on the treatment surface with the elbow flexed.
- Position the joint in its resting position.

Hand placement and fixation

- **Therapist's stable hand (right):** Grip the patient's distal arm from the anterior side and fixate it against the treatment surface; place your palpating finger in the radio-ulnar joint space.
- **Therapist's moving hand (left):** Hold the patient's distal forearm from the radial side; grip around the patient's distal radius.

Procedure

- Apply a Grade III distal glide movement to the radius in relation to the ulna by pulling and turning your body slightly to the left.

■ Mobilization progression (not shown)

- Position the elbow near the end range-of-motion into flexion or extension, pronation or supination.

Radio-ulnar joint proximal glide
for elbow and forearm hypomobility

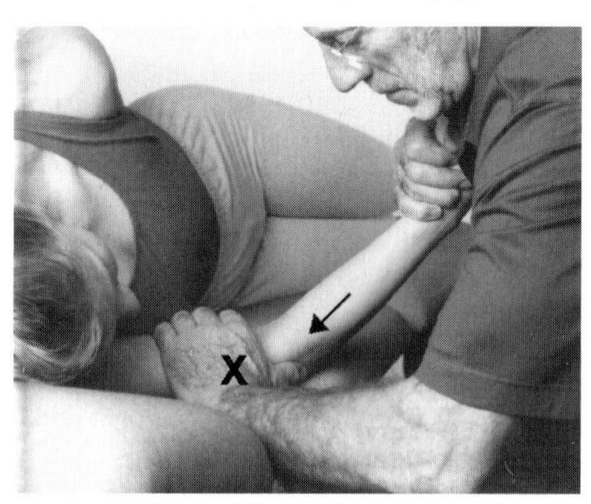

Figure 23 – mobilization in resting position

■ Figure 23: Mobilization in resting position

Objective

- To increase forearm pronation and supination, as well as elbow flexion and extension.
- To correct a distal positional fault of the radius in relation to the humerus by moving the radius proximally. See *Figure 20* for position testing technique.

Starting position

- The posterior side of the patient's arm and proximal ulna rests on the treatment surface with the elbow flexed.
- Position the joint in its resting position.

Hand placement and fixation

- **Therapist's stable hand (left):** Grip the patient's distal arm from the anterior side and fixate it against the treatment surface; place your palpating finger in the radio-ulnar joint space.
- **Therapist's moving hand (right):** Hold the patient's distal forearm from the radial side; grip around the patient's distal radius and thumb so that your thenar eminence is in contact with theirs; rest the patient's hand against your shoulder.

Procedure

- Apply a Grade III proximal glide movement to the radius in relation to the ulna by pushing along the long axis of the radius. Use your shoulder to apply additional force.

■ To correct a radius subluxation in children (not shown)

- Adapt the technique to correct a radius subluxation in children. Grip the child's wrist and forearm with a "handshake" grip; move the radius proximally while supinating or pronating the forearm.

■ Mobilization progression (not shown)

- Position the elbow near the end range-of-motion into flexion or extension, pronation or supination.

CHAPTER 12

ELBOW

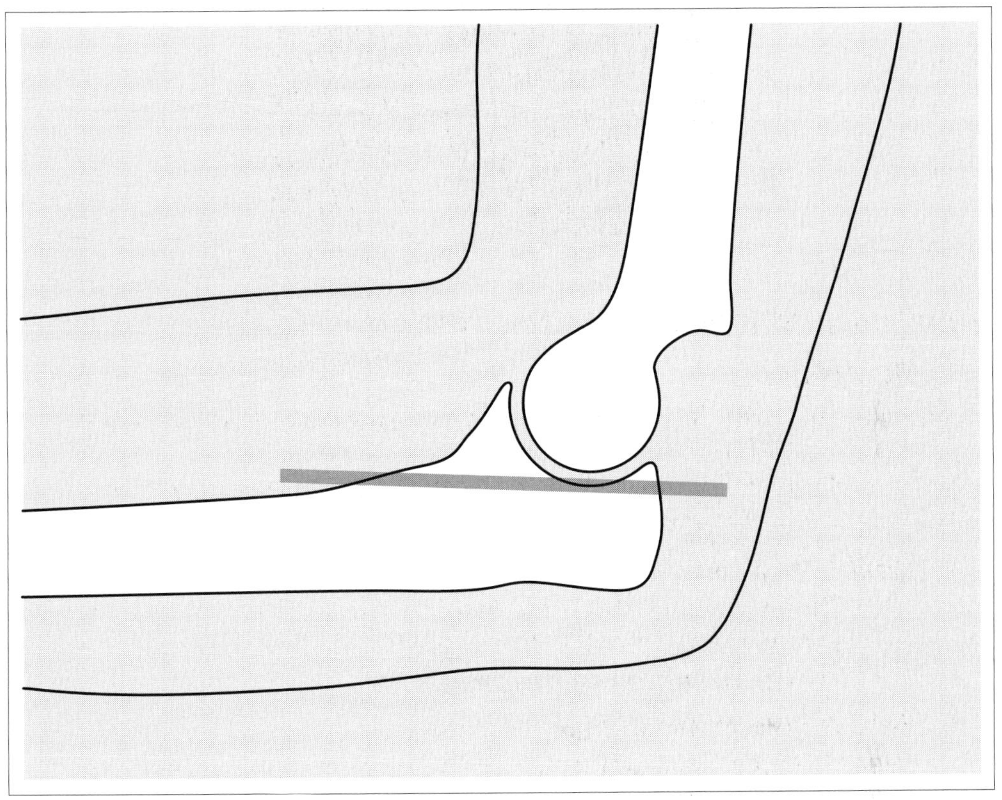

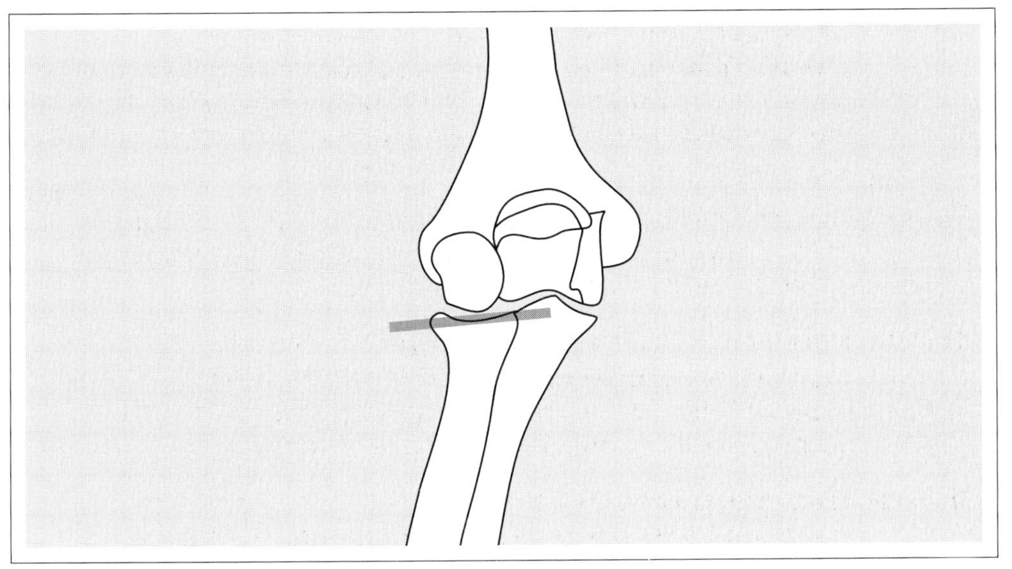

12 Elbow

Functional anatomy and movement

The elbow joint (art. cubiti) is an anatomically simple and mechanically compound joint and is divided into three joints:

Humero-ulnar joint
(art. humero-ulnaris)

The humero-ulnar joint is a biaxial saddle joint. The proximal thickened, wrench-shaped, concave trochlear notch of the ulna moves on the trochlea of the humerus.

Humeroradial joint
(art. humeroradialis)

The humeroradial joint belongs anatomically to the elbow, but functionally to the forearm and is described there (See *Chapter 11: Forearm*).

Proximal radio-ulnar joint
(art. radio-ulnaris proximalis)

The proximal radio-ulnar joint belongs anatomically to the elbow, but functionally to the forearm and is described there (See *Chapter 11: Forearm*).

Bony palpation

- Humerus, including the lateral supracondylar crest and lateral epicondyle
- Capitulum of the humerus
- Medial epicondyle
- Olecranon fossa
- Olecranon
- Humero-ulnar joint space
- Ulnar tuberosity
- Ulnar coronoid process

Ligaments

- Ulnar collateral ligament
- Radial collateral ligament

Bone movement and axes

- Flexion - extension: The transverse axis for flexion and extension passes through the trochlea and lies slightly oblique to the longitudinal axis of the upper arm. Therefore, in full extension the lower arm deviates laterally forming the valgus, or carrying angle, at the elbow which varies from 7° to 20°. Hyperextension of 5° to 15° is normal in children and many women because their olecranon process is smaller.
- Abduction - adduction: The humero-ulnar joint is a saddle joint. Therefore, with the elbow in flexion, slight abduction and adduction of the ulna is possible around a dorsal-ventral axis, which passes through the proximal part of the ulna. Abduction and adduction can only be performed passively.

End feel

- Flexion: hard end-feel. Flexion is limited by bone against bone, when the coronoid process of the ulna contacts the coronoid fossa of the humerus. (Active flexion is stopped by approximation of soft tissues on the ventral side of the upper arm and forearm.)
- Extension: hard end-feel. Extension stops when the olecranon of the humerus contacts the olecranon fossa of the ulna.

Joint movement (gliding)

- Ulna-humerus: Concave Rule
- Radius-humerus: Concave Rule

Treatment plane

- On the concave surface of the ulna
- On the concave surface of the radius

Zero position

- Both the upper arm and forearm lie in the frontal plane with the forearm supinated and the elbow extended

Resting position

- Humero-ulnar joint: elbow joint flexed approximately 70° and the forearm supinated approximately 10°
- Humeroradial joint: the elbow extended and the forearm fully supinated

Close-packed position

- Humero-ulnar joint: the elbow extended and the forearm supinated
- Humeroradial joint: the elbow flexed approximately 90° and the forearm supinated approximately 5°

Capsular pattern

- Flexion - extension. The proportion of these limitations is such that with flexion limited to 90° there is only 10° of limited extension.

Elbow examination scheme

(Refer to Chapters 3 and 4 for more information on examination)

Tests of function

A. Active and passive movements, including stability tests and end-feel

Flexion	150°
Extension	0° - 15° from zero

B. Translatoric joint play movements, including end-feel

Traction - compression	(Figure 24a)
Gliding	
Radial	(Figure 26)
Ulnar	(Figure 27)

C. Resisted movements	OTHER FUNCTIONS
Flexion	
Biceps brachii	Supination
Brachialis	
Brachioradialis	(see Forearm examination scheme)
Extension	
Triceps brachii, long head	Shoulder adduction, extension
Anconeus	

D. Passive soft tissue movements

Physiological

Accessory

E. Additional tests

Trial treatment

Traction	(Figure 24b)

Elbow

Elbow techniques

Elbow traction

for pain and hypomobility

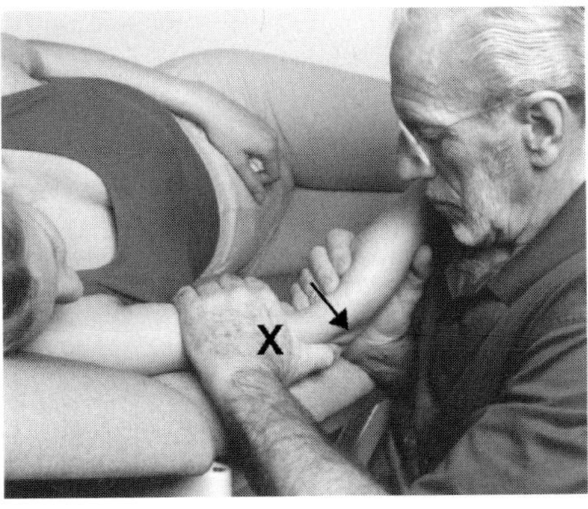

Figure 24a – test and mobilization in resting position

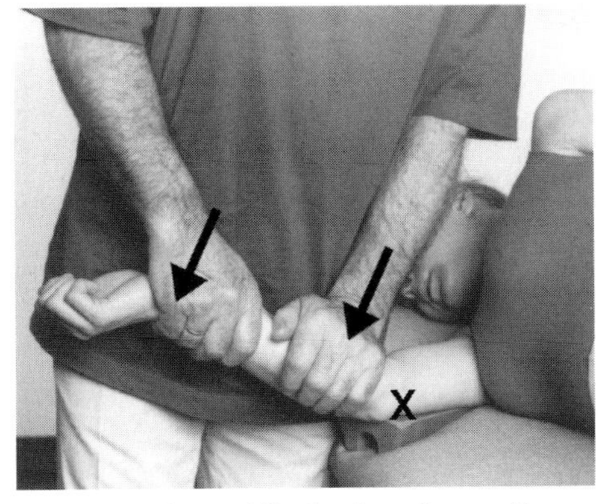

Figure 24b – mobilization in resting position

■ Figure 24a: Test and mobilization in resting position

Objective

- To evaluate the quantity and quality of posterior traction joint play in the humero-ulnar joint, including end-feel.
- To decrease pain or increase range-of-motion in the humero-ulnar joint.

Starting position

- The posterior side of the patient's arm rests on the treatment surface.
- Position the joint in its resting position.

Hand placement and fixation

- **Therapist's stable hand (left):** Grip the patient's distal arm from the anterior side and fixate it against the treatment surface; place your palpating finger in the humero-ulna joint space.
- **Therapist's moving hand (right):** Hold the patient's proximal forearm from the ulnar side; grip around the patient's proximal ulna.

Procedure

- Apply a Grade I, II, or III posterior traction movement to the ulna, at approximately a right angle to the forearm.

■ Figure 24b: Mobilization in resting position

- The dorsal side of the patient's humerus rests on the treatment surface.
- Enhance the fixation of the humerus by using a wedge under the humerus, and by positioning the patient in side-lying and/or by using a strap.
- Grip the patient's forearm from the ulnar side with both of your hands and your hypothenar eminence just distal to the joint space; hold the patient's forearm against your body; apply a Grade III posterior traction movement by bending your knees and leaning through your extended left arm.

Elbow traction
for restricted flexion and extension

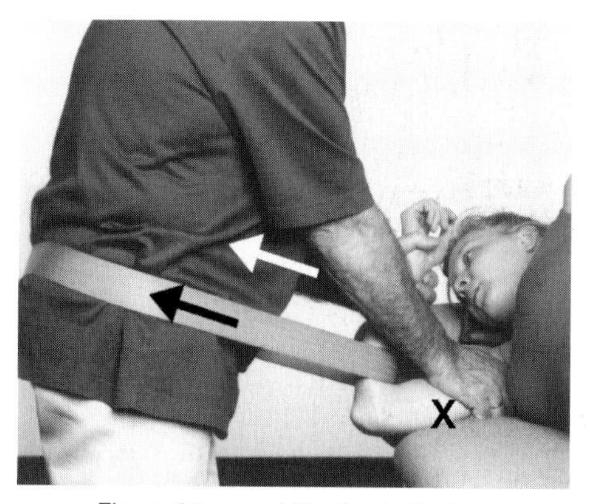

Figure 25a – mobilization in flexion

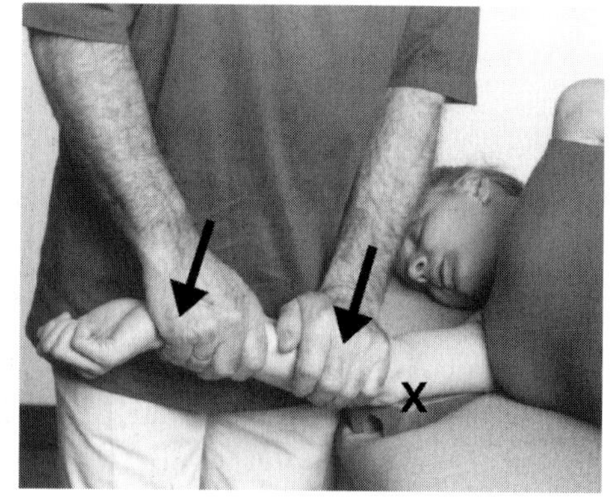

Figure 25b – mobilization in extension

■ Figure 25a: Flexion progression

Objective

- To increase flexion range-of-motion in the humero-ulnar joint.

Starting position

- The posterior side of the patient's humerus rests on the treatment surface; use a strap and position the patient side-lying to improve humerus fixation.
- Position the joint close to its end range-of-motion in flexion.

Hand placement and fixation

- **Therapist's stable hand (right)**: Fixate the patient's humerus against the treatment surface.
- **Therapist's moving hand (left):** Place a strap around the proximal forearm just distal to the joint space and around your body; hold the patient's distal forearm in your hand.

Procedure

- Apply a Grade III posterior/distal traction movement to the ulna at approximately a right angle to the forearm, by shifting your body backwards; move your body and your left hand as one.

■ Figure 25b: Extension progression

- The dorsal side of the patient's humerus rests on the treatment surface.
- Enhance the fixation of the humerus by using a wedge under the humerus, and by positioning the patient in side-lying and/or by using a strap.
- Grip the patient's forearm from the ulnar side with both of your hands and your hypothenar eminence just distal to the joint space; hold the patient's forearm against your body; apply a Grade III posterior traction movement by bending your knees and leaning through your extended left arm.
- Position the joint close to its end range-of-motion in extension.

Elbow radial glide

for restricted flexion and extension

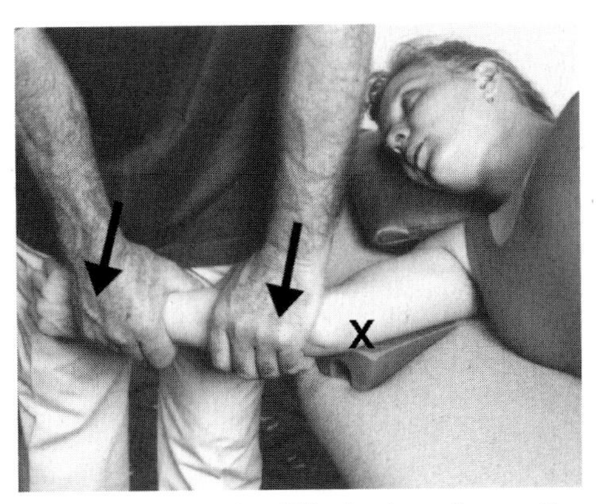

Figure 26 – test and mobilization in resting position

■ Figure 26: Test and mobilization in resting position

Objective

- To evaluate the quantity and quality of radial glide movement in the humero-ulnar joint, including end-feel.
- To increase elbow flexion and extension.

Starting position

- The lateral side of the patient's humerus rests on the treatment surface or wedge.
- Position the joint in its resting position.

Hand placement and fixation

- **Fixation:** The humerus is fixated against the treatment surface.
- **Therapist's moving hands:** Grip the patient's forearm from the ulnar side with both of your hands and your hypothenar eminence just distal to the joint space; hold the patient's forearm against your body.

Procedure

- Apply a Grade II or III radial glide movement to the ulna by bending your knees and leaning through your extended left arm.

■ Flexion and extension progression (not shown)

- Apply a Grade III radial glide movement to the ulna with the joint positioned close to its end range-of-motion in flexion or extension.

Elbow ulnar glide
for restricted flexion and extension

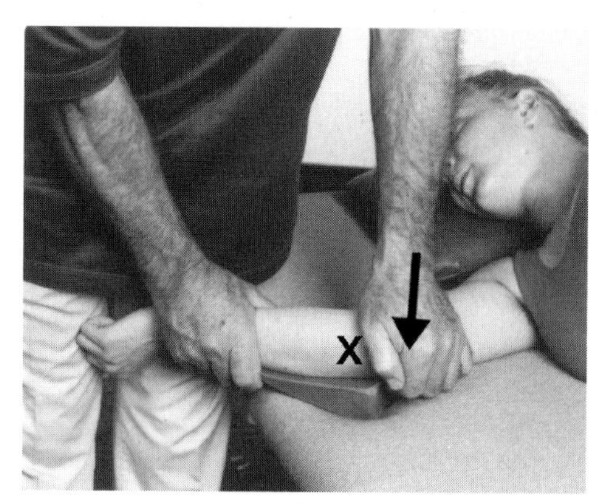

Figure 27 – test and mobilization in resting position

■ Figure 27: Test and mobilization in resting position

Objective

- To evaluate the quantity and quality of ulnar glide joint play in the humero-ulnar joint, including end-feel.
- To increase elbow flexion and extension.

Starting position

- The lateral side of the patient's proximal forearm and humerus contacts the treatment surface; place a wedge under the proximal forearm just distal to the joint space.
- Position the joint in its resting position.

Hand placement and fixation

- **Fixation:** The ulna is fixated against the wedge.
- **Therapist's stable hand (right)**: Hold the patient's distal forearm to support the position.
- **Therapist's moving hand (left):** Grip around the patient's distal humerus from the medial side with your thenar eminence just proximal to the joint space.

Procedure

- Apply a Grade II or III ulnar glide movement by moving the distal humerus in a radial direction.

■ Flexion and extension progression

- Apply a Grade III radial glide movement to the humerus with the joint positioned close to its end range-of-motion in flexion or extension.

CHAPTER 13

SHOULDER

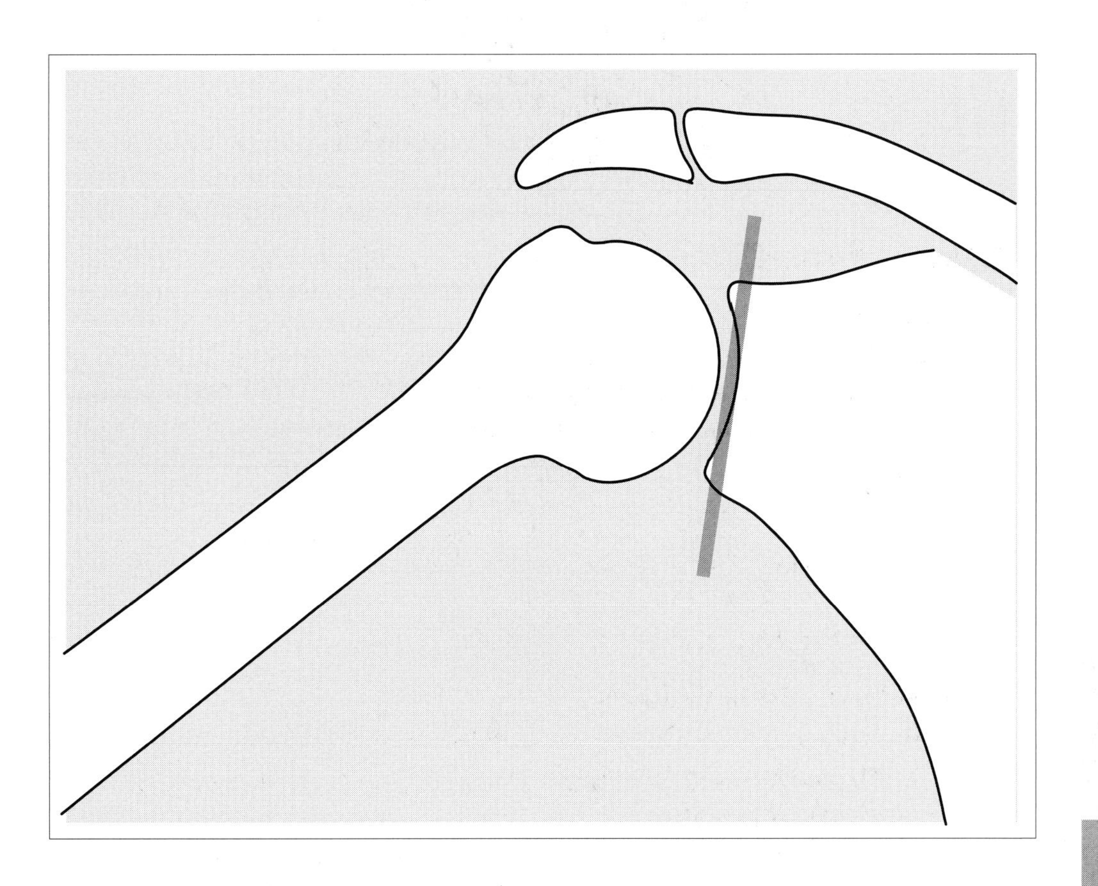

13 Shoulder

Functional anatomy and movement

Glenohumeral joint

(art. humeri)

The glenohumeral joint is an anatomically and mechanically simple triaxial joint (spheroid, unmodified ovoid). The convex surface of the head of the humerus articulates with the concave socket on the scapula.

The joint capsule of the glenohumeral joint is lax. When the arm hangs down in a dependent position, the medial side of the capsule folds loosely (recessus axillaris). This allows the shoulder joint a large range of movement. When the glenohumeral joint is immobilized for a long period of time, adhesions can form in these folds, and must be stretched for the shoulder to regain full mobility.

"Scapulohumeral rhythm" is described in *Chapter 14: Shoulder Girdle.*

Bony palpation

- Humeral head
- Lesser tubercle
- Deltoid tubercle
- Subacromial space
- Spine of scapula
- Supraspinous fossa
- Bicipital groove
- Greater tubercle
- Acromion (ventral, dorsal)
- Coracoid process
- Infraspinous fossa
- 1st rib
- Scapula, inferior angle, medial margin
- Scapula, superior angle, lateral margin

Ligaments

- Glenohumeral (superior, middle and inferior parts)
- Coracohumeral

Bone movement and axes

- Flexion - extension: around a transverse (medial-lateral) axis through the head of the humerus
- Abduction - adduction: around a sagittal (dorsal-ventral) axis through the head of the humerus
- Internal - external rotation: around a longitudinal axis through the length of the humerus

End feel

- Firm

Joint movement (gliding)

- Convex Rule

Treatment plane

- On the concave joint surface of the glenoid fossa of the scapula

Zero position

- The upper arm lies parallel to the trunk with the elbow extended and the thumb pointing ventrally.

Resting position

- Approximately 55° shoulder abduction, 30° horizontal adduction (i.e., the humerus lies in a vertical plane passing through the spine of scapula) and slight external rotation

Close-packed position

- Maximal abduction and lateral rotation

Capsular pattern

- Lateral rotation - abduction - medial rotation

■ Shoulder examination scheme

(Refer to Chapters 3 and 4 for more information on examination)

Tests of function

1. **Active and passive movements, including stability tests and end-feel**

With scapula fixated:

Movement in the sagittal plane around a transverse axis

Flexion	65°
Extension	35°

Movement in the frontal plane around a dorsal-ventral axis

Abduction	90°, 120° with lateral rotation
Adduction	8°

Movement in the transverse plane around a longitudinal axis

Horizontal adduction	30°
Medial rotation	90°
Lateral rotation	60°

2. **Translatoric joint play movements, including end-feel**

Traction - compression	(Figure 28a)
Gliding	
Caudal	(Figure 30a)
Ventral	(Figure 33)
Dorsal	(Figure 35)

3. **Resisted movements**

	OTHER FUNCTIONS
Flexion	
Coracobrachialis	Adduction
Deltoid	Abduction
Pectoralis major	Adduction, horizontal adduction
Biceps brachii	Adduction, elbow flexion
Extension	
Latissimus dorsi	Adduction, internal rotation, shoulder girdle depression (caudal)
Teres major	Adduction, internal rotation
Deltoid	Abduction, horizontal adduction
Triceps brachii	Adduction, elbow extension

Abduction	
Deltoid	Flexion, extension, abduction, horizontal adduction
Supraspinatus	
Adduction	
Teres minor	Extension, lateral rotation
Latissimus dorsi	Extension, medial rotation, shoulder girdle depression (caudal)
Teres major	Extension, medial rotation
Pectoralis major	Internal rotation, horizontal adduction
Lateral rotation	
Teres minor	Adduction
Infraspinatus	
Supraspinatus	Abduction
Medial rotation	
Subscapularis	
Pectoralis major	Horizontal adduction
Latissimus dorsi	Adduction, shoulder girdle depression (caudal)
Teres major	Adduction
Elbow Flexion	
Biceps brachii long head	Can cause pain with humeral rotation
Elbow Extension	
Triceps brachii	May cause pain with shoulder adduction and extension

4. Passive soft tissue movements

Physiological

Accessory

5. Additional tests

Trial treatment

Traction (Figure 28b)

Shoulder

■ Shoulder techniques

Shoulder traction
for pain and hypomobility

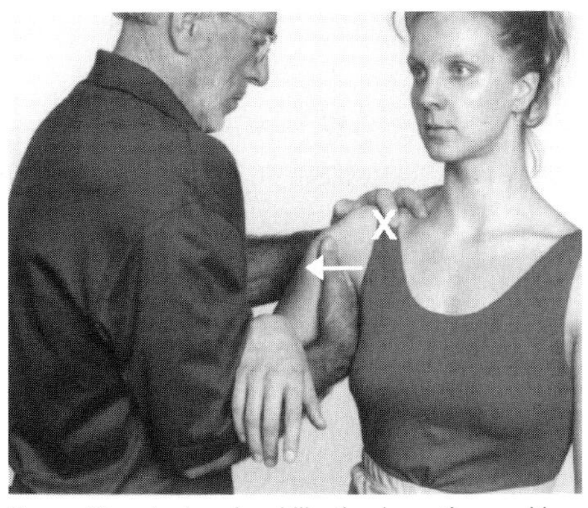

Figure 28a – test and mobilization in resting position

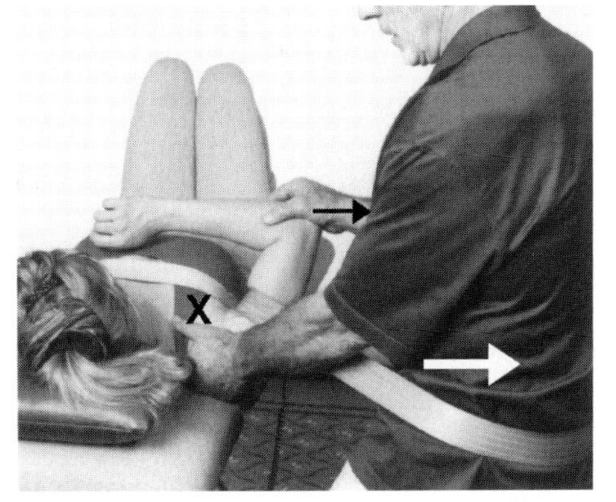

Figure 28b – mobilization in resting position

■ Figure 28a: Test and mobilization in resting position

Objective

- To evaluate the quantity and quality of traction joint play in the glenohumeral joint, including end-feel.
- To decrease pain or increase range-of-motion in the glenohumeral joint.

Starting position

- The patient sits with their forearm resting on your right forearm.
- The shoulder is positioned in its resting position.

Hand placement and fixation

- **Therapist's stable hand (left):** Grip the patient's shoulder from the posterior-superior side; place your palpating finger (left thumb shown) in the glenohumeral joint space.
- **Therapist's moving hand (right):** Grip the patient's proximal humerus from the medial side.

Procedure

- Apply a Grade I, II, or III traction movement to the glenohumeral joint with a lateral movement of the humerus.

■ Figure 28b: Mobilization in resting position

- The patient lies supine with the shoulder in the resting position and the elbow at approximately 90° of flexion.
- Fixate the patient's thorax and scapula to the treatment surface with a strap.
- Place a strap around the proximal humerus just distal to the joint space and around your body; grip the humerus with both your hands and support it in its resting position; apply a Grade III traction movement by leaning backward.

Shoulder traction
for restricted flexion and extension

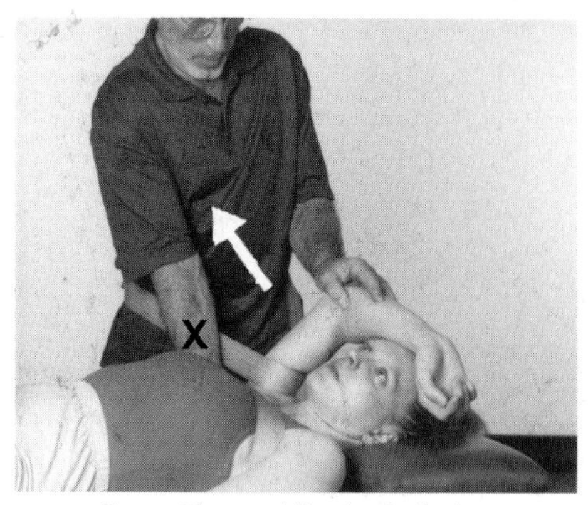

Figure 29a – mobilization in flexion

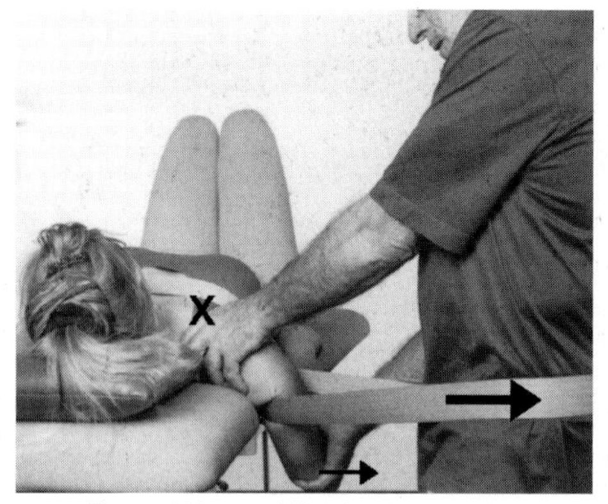

Figure 29b – mobilization in extension

■ Figure 29a: Flexion progression

Objective

- To increase flexion range-of-motion in the glenohumeral joint.

Starting position

- The patient lies supine with the shoulder close to the end range-of-motion into flexion and the elbow at approximately 90° flexion.

Hand placement and fixation

- **Fixation and therapist's stable hand (right)**: Fixate the patient's thorax and scapula to the treatment surface with a strap; enhance the fixation by pressing your right hand against the lateral border of the scapula.
- **Therapist's moving hand (left):** Place a strap around the proximal humerus just distal to the joint space and around your body; support the shoulder position by holding the arm with your hand.

Procedure

- Apply a Grade III traction movement by leaning backward; move your body and your left hand together as one.

■ Figure 29b: Extension progression

- Apply a Grade III traction movement with the patient's shoulder positioned near its end range-of-motion in extension.

Shoulder caudal glide
for restricted abduction

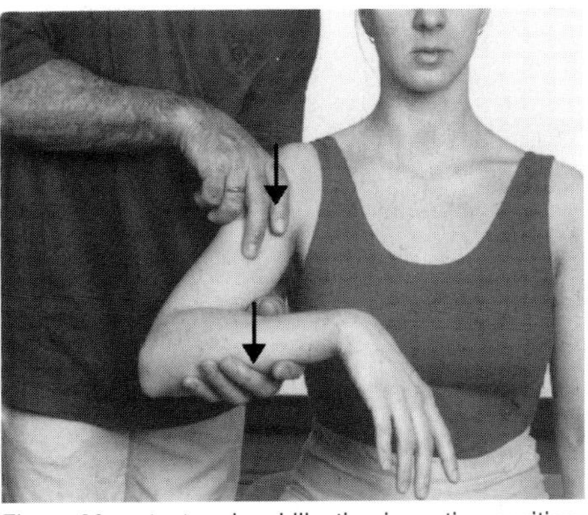

Figure 30a – test and mobilization in resting position

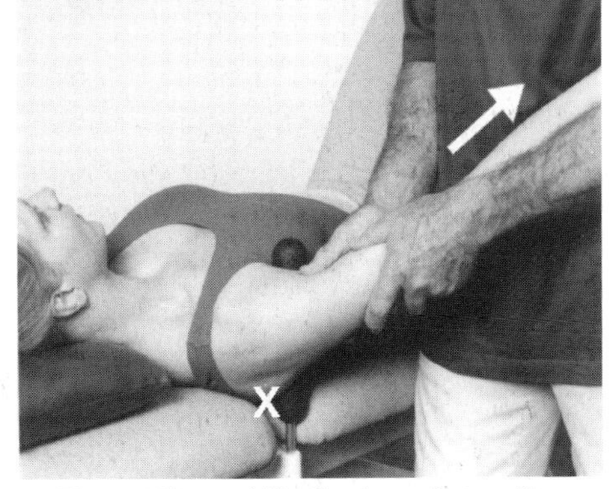

Figure 30b – mobilization in resting position

■ Figure 30a: Test and mobilization in resting position

Objective

- To evaluate the quantity and quality of caudal glide joint play in the glenohumeral joint, including end-feel.
- To increase shoulder abduction range-of-motion with increased humerus caudal glide joint play (Convex Rule).

Starting position

- The patient sits with their forearm resting on your left arm.
- The shoulder is positioned in its resting position.

Hand placement and fixation

- **Fixation:** No external fixation is required.
- **Therapist's moving hands:** Place your right hand on the head of the humerus just distal to the joint space; with your left hand, support the position of the patient's arm.

Procedure

- Apply a Grade II or III caudal glide movement to the glenohumeral joint by pressing down on the head of the humerus with your right hand; your right and left hand move together as one.

■ Figure 30b: Mobilization in resting position

- The patient lies supine with the shoulder in the resting position.
- Fixate the patient's scapula from the axilla with a pommel or stirrup attached to the treatment surface; if necessary, use an additional fixating strap around the patient's chest.
- Grip the humerus with both your hands and support it against your body; apply a Grade III traction movement by shifting your body backward.

■ Alternate technique: See Figure 31

Shoulder caudal glide
for restricted abduction (cont'd)

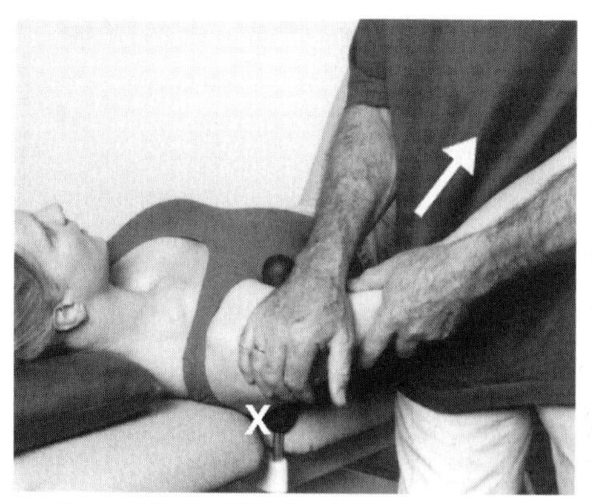

Figure 31 – mobilization in resting position

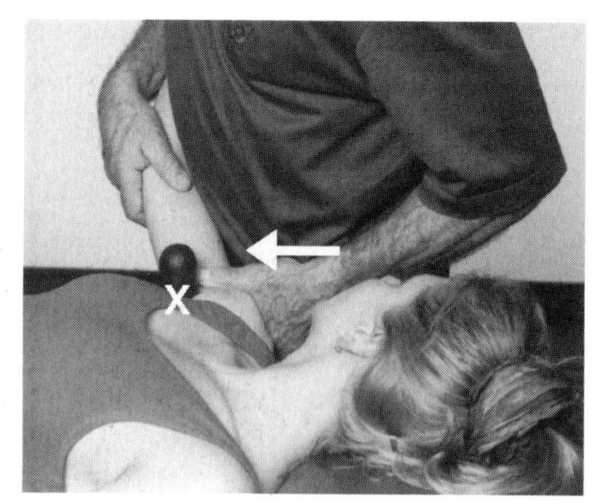

Figure 32 – mobilization in abduction

■ Figure 31: Mobilization in resting position

Objective

- To increase shoulder abduction by increasing humerus caudal glide joint play (Convex Rule).

Starting position

- The patient lies supine.
- Position the joint in its resting position.

Hand placement and fixation

- **Fixation:** Fixate the patient's scapula, from the axilla, with a pommel or stirrup attached to the treatment surface; if necessary, use an additional fixating strap around the patient's chest.
- **Therapist's moving hands:** Place your right hand on the head of the humerus just distal to the joint space; with your left hand and body, support the position of the patient's arm.

Procedure

- Apply a Grade III caudal glide movement to the glenohumeral joint by shifting your body backward while you press the head of the humerus in a caudal direction with your right hand; your hands and body move together as one.

■ Figure 32: Abduction progression

- The shoulder is positioned near the end range-of-motion into abduction.
- Adapt the same technique, but grip from the cranial side of the humerus; apply a Grade III caudal glide movement by shifting your body forward (Convex Rule).

Shoulder ventral glide
for restricted extension

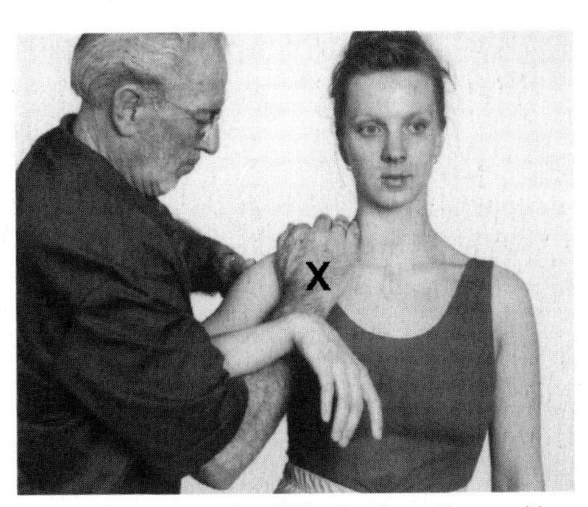

Figure 33 – test and mobilization in resting position

■ Figure 33: Test and mobilization in resting position

Objective

- To evaluate the quantity and quality of ventral glide joint play in the glenohumeral joint, including end-feel.
- To increase shoulder extension range-of-motion with increased humerus ventral glide joint play (Convex Rule).

Starting position

- The patient sits with their forearm resting on your right forearm.
- The shoulder is positioned in its resting position.

Hand placement and fixation

- **Therapist's stable hand (right)**: Fixate the patient's scapula from the ventral side, with firm pressure in the area of the acromion and corocoid process.
- **Therapist's moving hand (left):** Grip around the patient's arm from the dorsal side just distal to the joint space.

Procedure

- Apply a Grade II or III ventral glide movement to the humerus.

Shoulder ventral glide
for restricted extension-external rotation

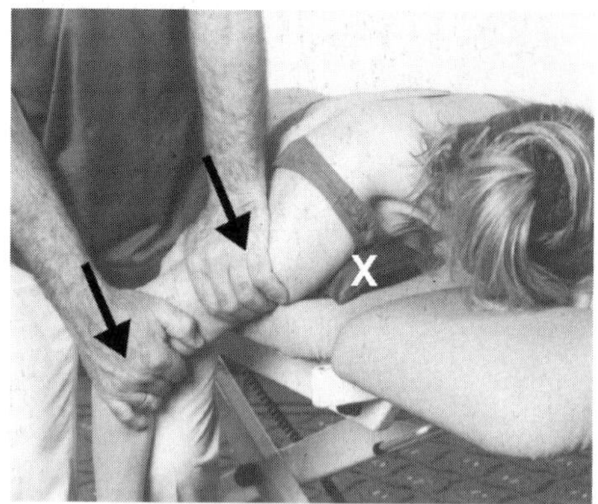

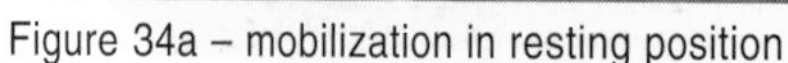
Figure 34a – mobilization in resting position

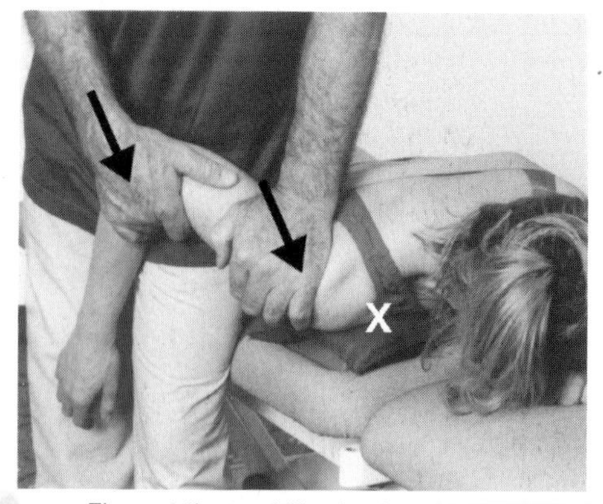

Figure 34b – mobilization in extension

■ Figure 34a: Mobilization in resting position

Objective

- To increase shoulder extension and external rotation with increased humerus ventral glide joint play (Convex Rule).

Starting position

- The patient lies prone with a wedge under the corocoid process and the arm beyond the edge of the treatment surface.
- Position the joint in its resting position.

Hand placement and fixation

- **Fixation:** The scapula is fixated by the wedge.
- **Therapist's moving hands:** Hold the patient's humerus against your body with both hands; grip with your left hypothenar eminence near the humeral head just distal to the joint space.

Procedure

- Apply a Grade III ventral glide movement to the glenohumeral joint by bending your knees and leaning through your extended left arm; move your hands and body together as one.

■ Figure 34b: Extension progression

- Apply a Grade III ventral glide movement to the glenohumeral joint with the shoulder positioned near its end range-of-motion in extension and external rotation.

Shoulder dorsal glide

for restricted flexion-internal rotation

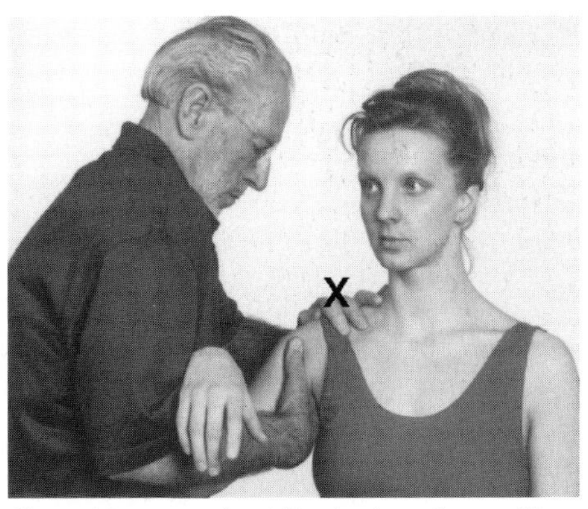
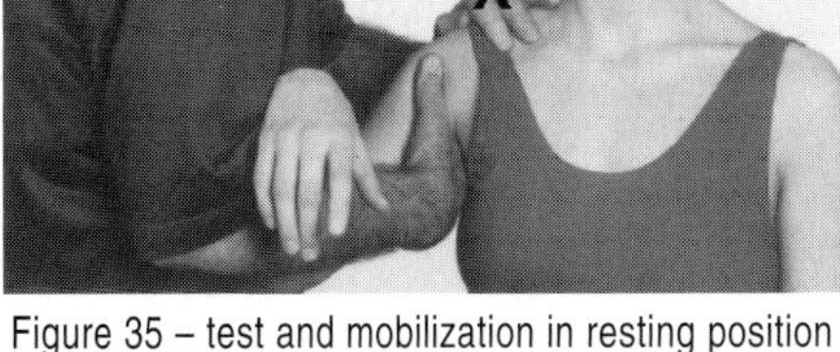

Figure 35 – test and mobilization in resting position

Figure 36 – mobilization in resting position

■ Figure 35: Test and mobilization in resting position

Objective

- To evaluate the quantity and quality of dorsal glide joint play in the glenohumeral joint, including end-feel.
- To increase shoulder flexion and internal rotation with increased humerus dorsal glide joint play (Convex Rule).

Starting position

- The patient sits with their forearm resting on your right forearm.
- The shoulder is positioned in its resting position.

Hand placement and fixation

- **Therapist's stable hand (left)**: Fixate the patient's scapula from the dorsal side with your index finger over the acromion.
- **Therapist's moving hand (right):** Grip around the patient's arm from the medial side just distal to the joint space.

Procedure

- Apply a Grade II or III dorsal glide movement to the humerus.

■ Figure 36: Mobilization in resting position

- The patient lies supine; fixate the patient's scapula with a wedge; hold the patient's humerus against your body with both hands; grip with your right hypothenar eminence near the humeral head just distal to the joint space; apply a Grade III dorsal glide movement to the glenohumeral joint by bending your knees and leaning through your extended right arm; move your hands and body together as one.

■ Notes

CHAPTER 14

SHOULDER GIRDLE

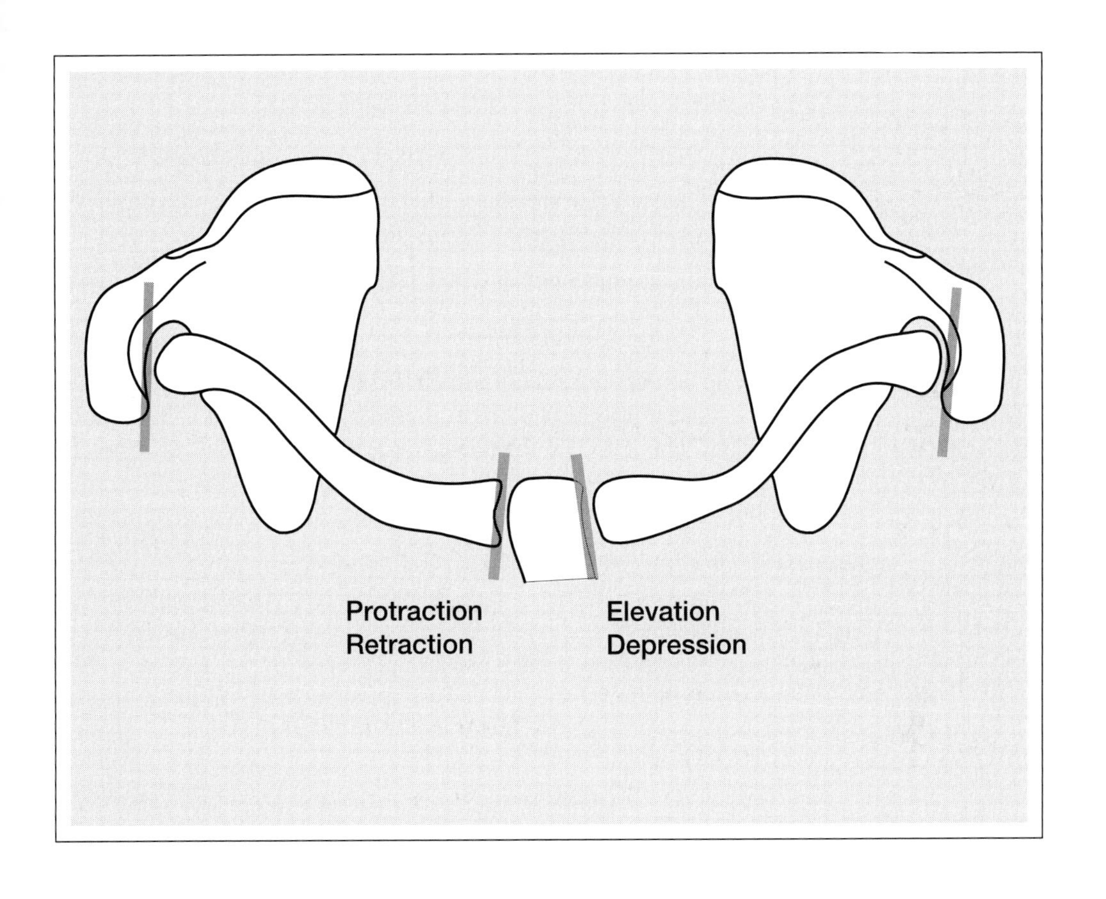

14 Shoulder girdle

(cingulum extremitatis superioris)

Functional anatomy and movement

The shoulder girdle consists of the scapula and clavicle with its connection to the trunk (manubrium sterni) with its joint facets, the clavicular notches (incisura clavicularis sterni).

The scapula is a triangular shaped bone with three borders:

- **The superior border** with the suprascapular notch and its superior transverse scapular ligament
- **The medial border** which is the longest of the three borders
- **The lateral border** which is the thickest border

These borders meet and form three angles: the **superior**, **inferior,** and **lateral angles** with the glenoid fossa located at the lateral angle. The spine of the scapula continues laterally to form the acromion which contains an articular facet for the clavicle cranially and the acromial angle caudally. The coraco-acromial ligament completes the coraco-acromial arch.

The medial or sternal end of the **clavicle** has a facet for articulation with the clavicular notch on the manubrium sterni. The lateral or acromial end has a facet for articulation with the acromion. The clavicle holds the scapula at the required distance from the thorax.

Arm and shoulder girdle movement

The word elevation is used to describe arm movement above the horizontal position. Movement of the entire shoulder girdle (scapula and humerus) enables the arm to elevate, producing flexion or abduction of the upper extremity. The scapula and humerus move together in a coordinated manner to produce what is called the "scapulo-humeral rhythm." Approximately two-thirds of this movement takes place at the glenohumeral joint and the remainder by movement of the scapula. Usually, movement is initiated at the glenohumeral joint and is followed by movement in the other joints. The ratio of humeral to scapular movement is 2:1 and takes place during both abduction and flexion.

At approximately 90° abduction, the greater tubercle of the humerus approaches the coraco-acromial ligament, preventing further movement. The humerus must laterally rotate so the greater tubercle can move dorsally under the coraco-acromial arch - this allows abduction beyond 90°.

For maximal elevation of the arm, abduction and external rotation of the scapula, elevation and rotation of the clavicle, and flattening of the thoracic kyphosis are all necessary. Therefore, full elevation of the arm requires that many joints function normally: 1) gleno-humeral joint, 2) acromioclavicular joint, 3) sternoclavicular joint, 4) thoracic spine joints, including rib articulations, and 5) the scapulo-thoracic "joint."

The shoulder girdle has two joints: the sternoclavicular joint and the acromioclavicular joint.

■ The sternoclavicular joint

(art. sternoclavicularis, abbreviated SC)

The sternoclavicular joint is an anatomically compound and mechanically simple biaxial joint (sellaris, unmodified sellar). An articular disc divides the joint cavity into two parts. It technically is considered a saddle joint (sellaris), but because there is a lax capsule and the disc is flexible, it is functionally a triaxial joint (spheroid, unmodified ovoid).

Bony palpation

- Sternum
- Medial clavicle
- Sternoclavicular joint space
- Infraclavicular fossa

Ligaments

- Sternoclavicular ligaments (anterior and posterior)
- Interclavicular ligament
- Costoclavicular ligament

Bone movement and axes

For convenience in describing clavicular movements, the clavicle is considered moving on a stationary manubrium sterni around three axes:

- Elevation – Depression: around the sagittal (dorso-ventral) axis through the medial end of the clavicle. The clavicle moves with its convex surface around the sagittal axis for cranial and caudal movements.
- Protraction – Retraction: around the vertical (cranial-caudal) axis which passes longitudinally through the manubrium sterni. The clavicle's concave surface along with the disc moves around the vertical axis for ventral and dorsal movements.

- Rotation (with shoulder flexion - extension): around the longitudinal axis which passes lengthwise through the clavicle. The anterior border of the clavicle moves cranially (lateral rotation of the clavicle) during shoulder flexion, and caudally (medial rotation of the clavicle) during shoulder extension.

End feel

- Unknown

Joint movement (gliding)

- Cranial-caudal movement: Convex Rule
- Dorsal-ventral movement: Concave Rule

Treatment plane

- Elevation - depression: on the concave joint surface of the sternum
- Protraction - retraction: on the concave joint surface of the clavicle

Zero position

- See *Shoulder girdle zero and resting position* (page 161).

Resting position

- See *Shoulder girdle zero and resting position* (page 161).

Close-packed position

- Arm in full elevation

Capsular pattern

- Unknown

■ The acromioclavicular joint

(art. acromioclavicularis, abbreviated AC)

The acromioclavicular joint is an anatomically simple (or compound when a disc is present) and mechanically compound plane gliding joint. Anatomically the acromion is concave and the clavicle is convex. However, it is functionally a triaxial ball and socket joint (spheroidea) due to its lax capsule and a flexible disc which is usually present.

Bony palpation

- Acromion
- Lateral clavicle
- Acromioclavicular joint space

Ligaments

- Acromioclavicular ligament
- Coracoclavicular ligament (trapezoid and conoid parts)

Bone movement and axes

In describing movements of this joint, one can consider that the scapula moves in relation to a stationary clavicle around three axes. At the same time, the scapula also moves in relation to the thorax. There are three axes of movement for the acromioclavicular joint.

Scapula cranial - caudal: There is very little movement here. A sagittal (dorso-ventral) axis runs through the lateral end of the clavicle. *Figure 37a* (ventral view) shows the clavicular-scapular vertical angle (C-S V) which is approximately 90°. *Figure 37b* illustrates the axis (dark dot) around which the scapula moves. As the inferior angle of the scapula moves laterally, the C-S V angle increases and, as it moves medially, the angle decreases. As a result of these movements, the glenoid cavity will face more cranially (as the C-S V angle increases) or more caudally (as the C-S V angle decreases).

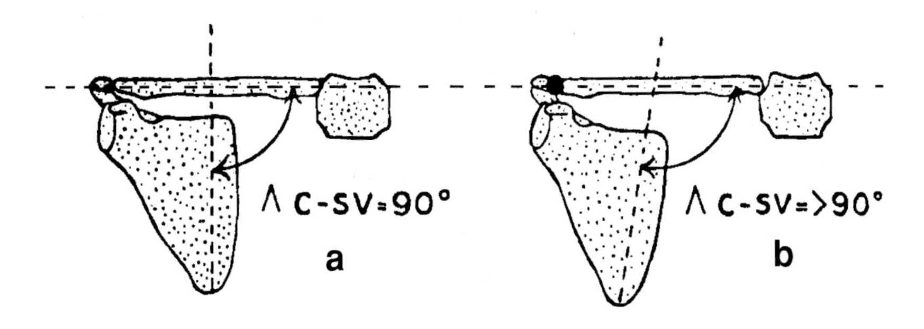

Figure 37
Axis of movement with change in the clavicular-scapular vertical angle (C-S V)

Scapula ventral - dorsal: A vertical (cranial-caudal) axis passing through the lateral end of the clavicle. *Figure 38a* (cranial view) illustrates the clavicular-scapular horizontal angle (C-S H) which is approximately 60°. *Figure 38b* shows the axis (dark dot) around which the scapula moves. When the medial border of the scapula moves away from the thorax ("winging" or abduction of the scapula), the C-S V angle increases (*Figure 38b*). As the medial border of the scapula moves towards the thorax (adduction), the angle decreases. As a result of these movements, the glenoid cavity will face more ventrally (with abduction) or more dorsally (with adduction).

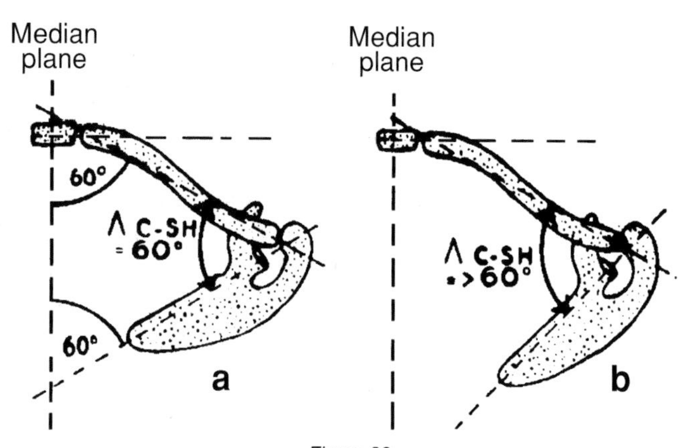

Figure 38
Axis of movement with change in the clavicular-scapular horizontal angle (C-S H)

- **Scapula rotation**: A longitudinal axis which passes lengthwise through the clavicle allows movements of the scapula over the thorax. During lateral rotation the inferior angle moves laterally and forward and the glenoid cavity then faces more cranially and ventrally. With medial rotation, the inferior angle moves medially and backwards, and therefore the glenoid cavity faces more caudally and dorsally.

End feel

- Unknown

Joint movement (gliding)

- Retraction - protraction
  Elevation - depression
- Apply the Concave Rule or Convex Rule according to whether the mobilization technique moves the concave or convex surface of the targeted bone.

Treatment plane

- Lies on the concave surface of the acromion

Zero position

- See *Shoulder girdle zero and resting position.*

Resting position

- See *Shoulder girdle zero and resting position.*

Close-packed position

- Arm in 90° abduction

Capsular pattern

- Unknown

The scapula is positioned more cranially and tightly against the thorax in muscular individuals. In less muscular individuals, the scapula is positioned less tightly against the thorax ("winged") and more caudally. Between these extremes of positions, the scapula lies with the superior angle at the level of the second rib, the inferior angle at the level of the seventh rib, and the medial border approximately five centimeters lateral to the spinous processes. A vertical plane through the scapula forms an angle of approximately 50° with the medial plane (Figure 38a).

The superior surface of the **clavicle** lies approximately in the horizontal plane and forms with the medial plane an angle of approximately 60° (Figure 38a).

The **clavicular-scapular horizontal angle** (C-S H) is therefore approximately 60° as it forms the third angle of an equilateral triangle (Figure 38a).

■ Shoulder girdle examination scheme

(Refer to Chapters 3 and 4 for more information on examination)

Tests of function

Sternoclavicular joint

1. **Active and passive movements, including stability tests and end-feel**

Shoulder girdle

Elevation	45°
Depression	7°
Protraction	30°
Retraction	20°

Clavicle rotation around its longitudinal axis during internal and external rotation of the abducted arm:

Lateral rotation	10°
Medial rotation	10°

2. **Translatoric joint play movements, including end-feel**

Traction - compression	(Figure 39a)
Gliding	
Cranial	(Figure 39d)
Caudal	(Figure 39e)
Ventral	(Figure 39f)
Dorsal	(Figure 39g)

3. **Resisted movements**

Shoulder girdle elevation
- *Trapezius*
- *Levator scapula*

Shoulder girdle depression
- *Lower trapezius*
- *Serratus anterior*
- *Latissimus dorsi*
- *Pectoralis minor*
- *Subclavius*

Shoulder girdle protraction
- *Pectoralis major*
- *Pectoralis minor*
- *Serratus anterior*

Shoulder girdle retraction
- *Trapezius*
- *Rhomboids*
- *Latissimus dorsi*

4. **Passive soft tissue movements**

Physiological

Accessory

5. **Additional tests**

Acromioclavicular joint

1. **Active and passive movements, including stability tests and end-feel**

   Scapula:

Lateral rotation	25°
Medial rotation	25°

2. **Translatoric joint play movements, including end-feel**

Traction - compression	(adapt from Figure 39a, fixate the clavicle, move the scapula)
Gliding	
Ventral - dorsal	(Figure 40a)
With the scapula	(Figure 41, 42)

3. **Resisted movements**

   Scapula lateral rotation

   Trapezius

   Serratus anterior

   Scapula medial rotation

   Rhomboids

   Levator scapula

4. **Passive soft tissue movements**

   Physiological

   Accessory

5. **Additional tests**

Trial treatment

Sternoclavicular joint	(Figure 39b)
Acromioclavicular joint	(adapt from Figure 39b)

Shoulder girdle techniques

Sternoclavicular joint

Acromioclavicular joint

Scapula

Clavicle-sternum traction
for pain and hypomobility

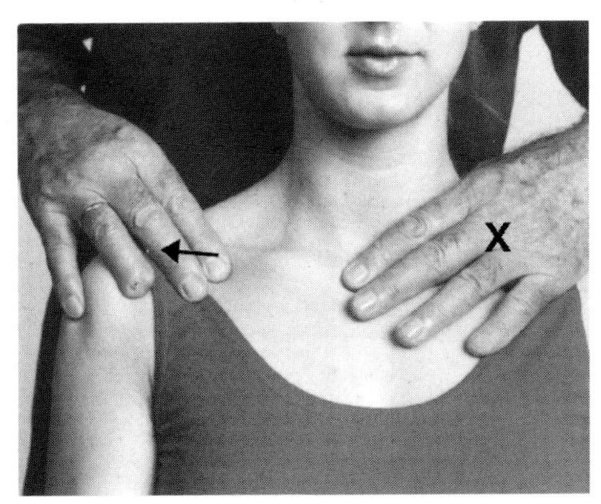

Figure 39a – test and mobilization in resting position

■ Figure 39a: Test and mobilization in resting position

Objective

- To evaluate the quantity and quality of traction joint play in the sternoclavicular joint, including end-feel.
- To decrease pain and increase shoulder girdle mobility with increased sternoclavicular traction joint play.

Starting position

- The patient sits. The sternoclavicular joint is in its resting position.

Hand placement and fixation

- **Therapist's stable hand (left)**: Fixate the patient's sternum and body against your body.
- **Therapist's moving hand (right):** Grip around the patient's clavicle from the ventral side.

Procedure

- Apply a Grade I, II, or III lateral movement to the clavicle to produce a traction movement in the sternoclavicular joint.

Clavicle-sternum traction

for hypomobility

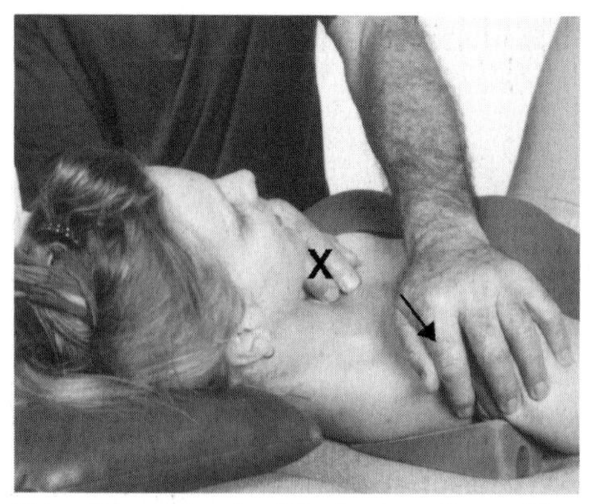

Figure 39b – test and mobilization in resting position

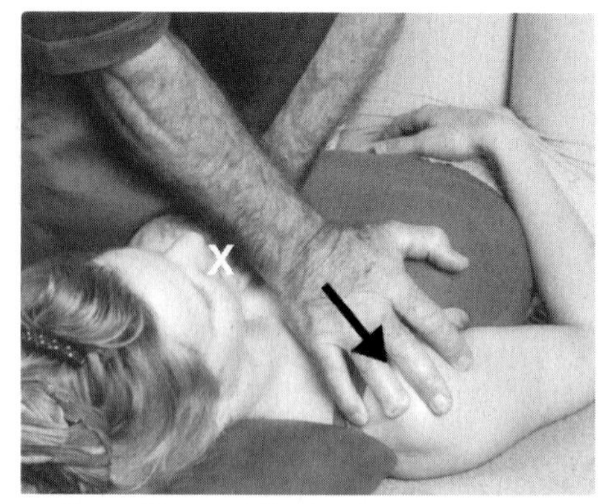

Figure 39c – mobilization in resting position

■ Figure 39b: Test and mobilization in resting position

Objective

- To evaluate the quantity and quality of traction joint play in the sternoclavicular joint, including end-feel.
- To increase shoulder girdle mobility with increased sternoclavicular traction joint play.

Starting position

- The patient lies supine with a sandbag under the scapula.
- Stand on the side of the patient opposite the targeted sternoclavicular joint.

Hand placement and fixation

- **Therapist's stable hand (right)**: Press the patient's sternum and thorax toward the treatment surface to fixate the sternum.
- **Therapist's moving hand (left):** Grip around the patients shoulder with your thenar eminence on the lateral clavicle.

Procedure

- Apply a Grade II or III traction movement by leaning through your extended left arm.

■ Figure 39c: Alternate mobilization technique

- Apply a Grade III traction movement using the same procedure and switching your left and right hand grips.

Clavicle-sternum cranial and caudal glide
for restricted depression and elevation

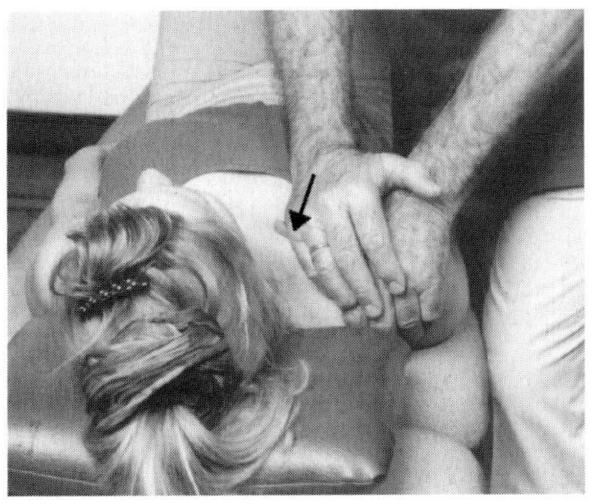

Figure 39d – test and mobilization, cranial

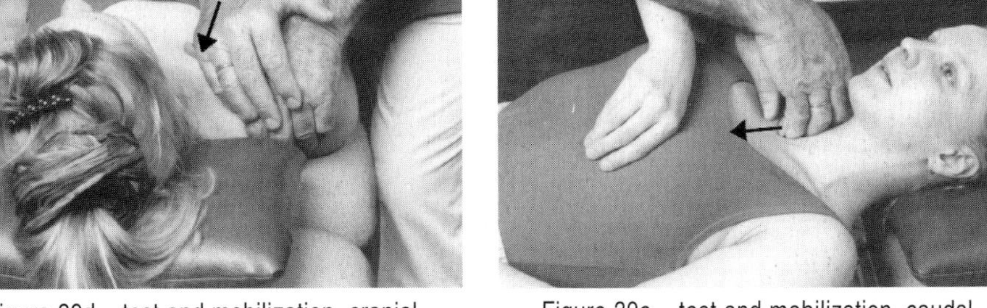

Figure 39e – test and mobilization, caudal

■ Figure 39d: Test and mobilization, cranial

Objective

- To evaluate the quantity and quality of cranial glide joint play in the sternoclavicular joint, including end-feel.
- To increase scapular depression range-of-motion with increased cranial glide joint play in the sternoclavicular joint (Convex Rule).

Starting position

- The patient lies supine.

Hand placement and fixation

- **Fixation**: No external fixation is necessary.
- **Therapist's moving hands:** Place the length of your left thumb and thenar eminence along the caudal surface of the patient's clavicle; place your right hypothenar eminence over your left thumb to reinforce your grip.

Procedure

- Lean through your extended arms to apply a Grade II or III cranial glide movement to the clavicle. When testing, palpate the joint space.

■ Figure 39e: Test and mobilization, caudal

- Apply a Grade III caudal glide movement to increase scapular elevation (Convex Rule). With your left hand, place your finger along the cranial surface of the patient's clavicle; reinforce your grip with the fingers of your right hand; shift your body backward and pull through your extended arms. When testing, palpate the joint space.

■ Mobilization progression (not shown)

- Position the shoulder girdle near its end range-of-motion in elevation or depression.

Clavicle-sternum ventral and dorsal glide

for restricted protraction and retraction

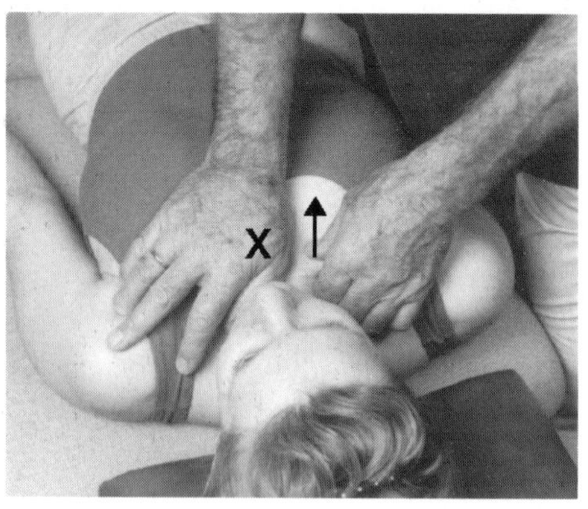

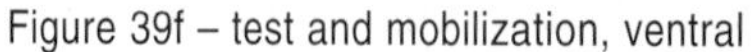

Figure 39f – test and mobilization, ventral

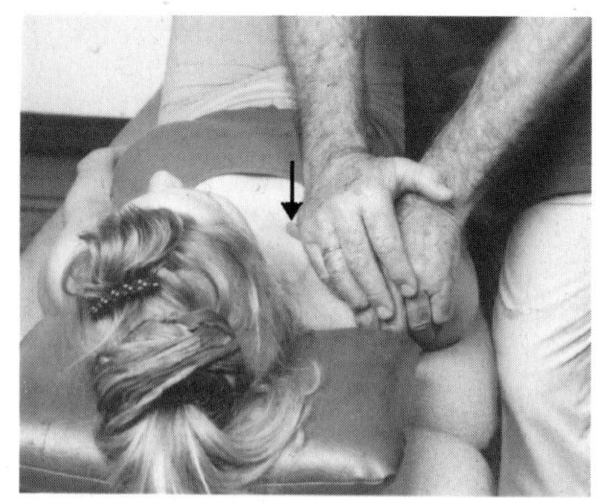

Figure 39g – test and mobilization, dorsal

■ Figure 39f: Test and mobilization, ventral

Objective

- To evaluate the quantity and quality of ventral glide joint play in the sternoclavicular joint, including end-feel.
- To increase scapular protraction range-of-motion by increasing ventral glide joint play in the sternoclavicular joint (Concave Rule).

Starting position

- The patient lies supine.

Hand placement and fixation

- **Therapist's stable hand (right)**: Fixate the superior aspect of the sternum with pressure from your thenar eminence.
- **Therapist's moving hand (left):** Grip around the clavicle with your fingers.

Procedure

- Lift the clavicle in an anterior direction to apply a Grade II or III ventral glide joint play movement (shown). When testing, use your right hand to palpate in the joint space rather than to fixate the sternum (not shown).

■ Figure 39g: Test and mobilization, dorsal

- Apply a Grade III dorsal glide movement to the clavicle to increase scapular retraction (Concave Rule). Place the length of your left thumb and thenar eminence along the ventral surface of the patient's clavicle; place your right hand over your left thumb to reinforce your grip; lean through your extended arms to apply the dorsal glide movement to the clavicle. When testing, palpate the joint space.

■ Mobilization progression (not shown)

- Position the shoulder girdle near its end range-of-motion into protraction or retraction.

Clavicle-acromion ventral glide

for hypomobility

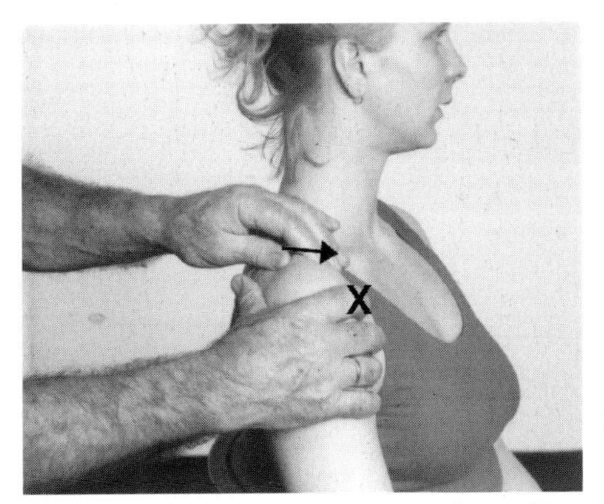

Figure 40a – test and mobilization in resting position

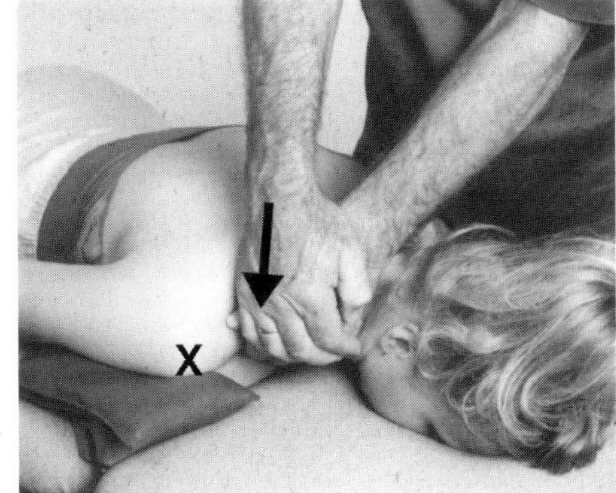

Figure 40b – mobilization in resting position

■ Figure 40a: Test and mobilization in resting position

Objective

- To evaluate the quantity and quality of ventral glide joint play in the acromioclavicular joint, including end-feel.
- To increase shoulder girdle mobility with increased ventral glide joint play in the acromioclavicular joint.

Starting position

- The patient sits. The acromioclavicular joint is in its resting position.

Hand placement and fixation

- **Therapist's stable hand (right)**: Fixate the patient's scapula by gripping the acromion or corocoid process from the ventral side, and the spine of the scapula from the dorsal side.
- **Therapist's moving hand (left):** Grip around the clavicle with your fingers just proximal to the joint space.

Procedure

- Press the clavicle in a ventral direction to apply a Grade II or III ventral glide movement (shown). When testing, use your right index finger to palpate in the joint space (not shown).

■ Figure 40b: Alternate mobilization technique

- Apply a Grade III ventral glide movement to the clavicle. The patient lies prone; fixate the scapula with a sandbag under the acromion, taking care that the sandbag does not contact the clavicle; place your left thumb and thenar eminence along the clavicle from the dorsal side; place your right hand over your left thumb to reinforce your grip; lean through your extended arms.

Scapula caudal glide and winging
for hypomobility

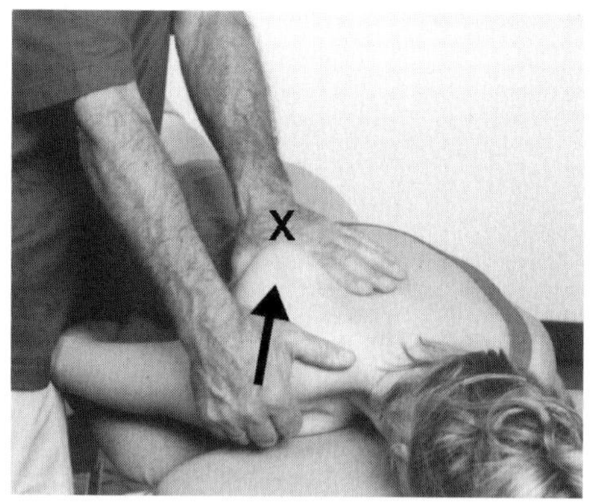

Figure 41 – mobilization caudal in resting position

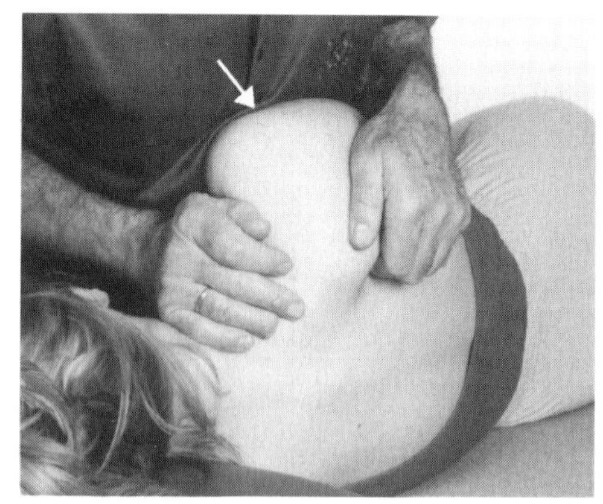
Figure 42 – mobilization winging in resting position

■ Figure 41: Mobilization caudal in resting position

Objective

- To increase shoulder girdle mobility by increasing caudal glide in the "scapulo-thoracic" articulation.
- To increase mobility in the sternoclavicular joint.

Starting position

- The patient is prone. The scapula is in its resting position.

Hand placement and fixation

- **Therapist's stable hand (left)**: Fixate the patient's scapula by placing your hand around the inferior angle of the scapula.
- **Therapist's moving hand (right):** Grip around the patient's scapula from the ventral side.

Procedure

- Lift the scapula in a dorsal-medial-caudal direction to apply a Grade III caudal glide joint play movement.

■ Figure 42: Mobilization winging in resting position

- The patient is side-lying; hold the patient's scapula against your body with both hands: your right hand grips the shoulder girdle from the cranial side, your left hand grips the inferior angle of the scapula; apply a Grade III winging mobilization by leaning your body downward on the patient's scapula and lifting the inferior border of the scapula laterally at the same time. This produces a winging movement in the scapula.

■ Alternate mobilization techniques (not shown)

- Use the grip described in Figure 42 to mobilize the scapula in cranial, caudal, medial, and lateral directions. These mobilizations also increase mobility in the sternoclavicular joint.
- Use the grip described in figure 42 to mobilize the scapula into internal and external rotation through the long axis of the clavicle. These mobilizations also increase mobility in the acromioclavicular joint.

■ Notes

CHAPTER 15

TOES

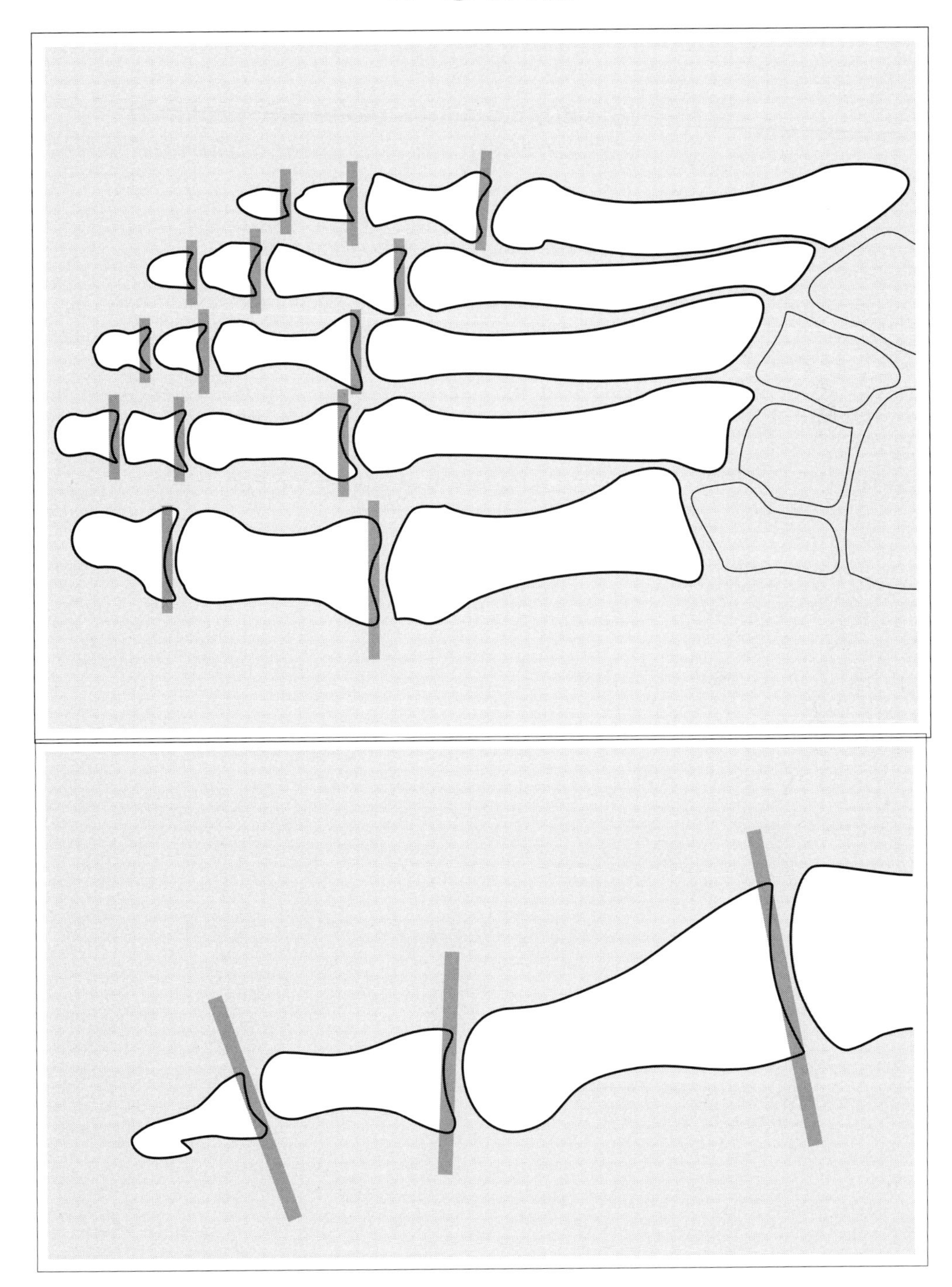

15 Toes

■ Functional anatomy and movement

■ Interphalangeal joints, distal and proximal

(artt. interphalangeals pedis, distalis et proximalis, abbreviated DIP and PIP)

■ Metatarsophalangel joints

(artt. metatarsophalangeae, abbreviated MTP)

The interphalangeal joints of the toe and the metatarsophalangeal joints are anatomically similar to the fingers. Each phalanx has a convex head distally and a concave base proximally. The axes of movement are the same as described for the fingers. Abduction and adduction of the toes around a dorsal-plantar axis results in movement away from and toward the second toe, which is regarded as lying in the middle of the foot.

Bony palpation

- Toe bones
- DIP and PIP joint spaces

Ligaments

- Collateral ligaments

Bone movement and axes

DIP and PIP:

- Flexion - extension: around a tibiofibular axis through the head of the phalanx

MTP:

- Tibial - fibular flexion: around a dorsalplantar axis through the head of the metatarsals. (Also called abduction and adduction movement away from and toward the 2nd toe.)
- Rotation (passive): around a longitudinal axis through the phalanx

End feel

- Firm

Joint movement (gliding)

- Concave Rule

Treatment plane

- Lies on the concave surface at the base of the phalanx.

Zero position

- The longitudinal axes through the metatarsals and corresponding phalangeal bones form a straight line.

Resting position

- *DIP and PIP:* slight flexion
- *MTP:* approximately 10° extension (from zero)

Close-packed position

- *DIP, PIP and MTP I:* maximal extension
- *MTP II-V:* maximal flexion

Capsular pattern

- Restricted in all directions, DIP and PIP extension more limited, MTP slightly more limited in flexion

■ Toe examination scheme

(Refer to Chapters 3 and 4 for more information on examination)

Tests of function

1. **Active and passive movements, including stability tests and end-feel**

DIP	Flexion	55°
PIP	Flexion	40°
MTP	Flexion	40°
	Extension from zero	40°
	Abduction	varies greatly with individuals

2. **Translatoric joint play movements, including end-feel**

Traction - compression	(Figure 43a)
Gliding	
Plantar	(Figure 44a)
Dorsal	(Figure 45a)
Tibial - fibular	(Figure 46a)

3. **Resisted Movements**

Flexion	ACTS ON:
Flexor digitorum brevis	PIP
Flexor digitorum longus	DIP
Flexor hallucis brevis	MTP
Flexor hallucis longus	IP
Flexor digiti minimi brevis	MTP
Lumbricals	MTP flexion; DIP, PIP extension
Extension	
Extensor digitorum	DIP, PIP
Extensor hallucis brevis	MTP
Extensor hallucis longus	IP
Lumbricals	DIP, PIP
Abduction	
Dorsal interossei	MTP
Abductor digiti minimi	MTP
Abductor hallucis	MTP
Adduction	
Adductor hallucis	MTP
Plantar interossei	MTP

4. **Passive soft tissue movements**

   Physiological

   Accessory

5. **Additional tests**

Trial treatment

Traction	(Figure 43b)

Toe techniques

Toe traction
for pain and hypomobility

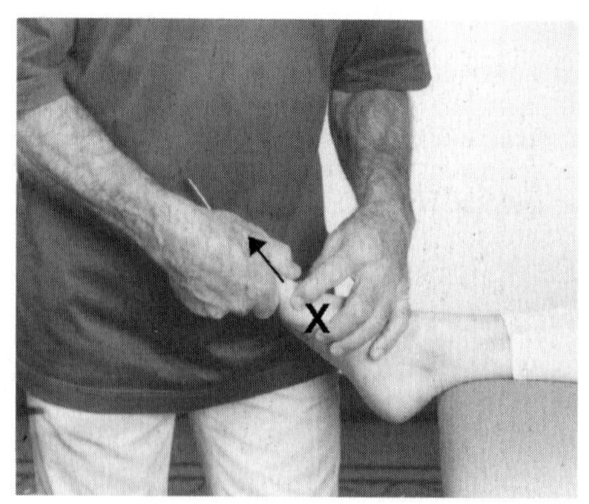

Figure 43a – test and mobilization in resting position

Figure 43b – mobilization in resting position

■ Figure 43a: Test and mobilization in resting position

Objective

- To evaluate the quantity and quality of traction joint play in a DIP, PIP, or MTP joint, including end-feel.
- To decrease pain or increase range of motion in a DIP, PIP, or MTP joint.

Starting position

- The plantar side of the patient's foot faces down.
- Position the targeted DIP, PIP, or MTP joint in its resting position.

Hand placement and fixation

- **Therapist's stable hand (left):** Hold the patient's foot in your hand; grip with your fingers just proximal to the targeted joint space; fixate the patient's foot against your body.
- **Therapist's moving hand (right):** Hold the patient's toe in your hand; grip with your fingers just distal to the targeted joint space

Procedure

- Apply a Grade I, II or III distal traction movement to the distal phalanx.

■ Figure 43b: Mobilization in resting position

- Improve your "handle" on the toe by gripping with tape, a paper towel, or a tongue depressor taped to the toe; apply a Grade III distal traction movement to the distal phalanx.

Toe traction
for restricted flexion and extension

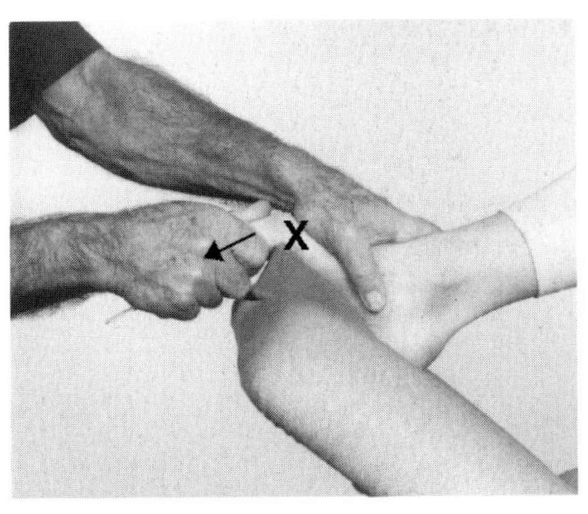

Figure 43c – MCP traction-mobilization in flexion

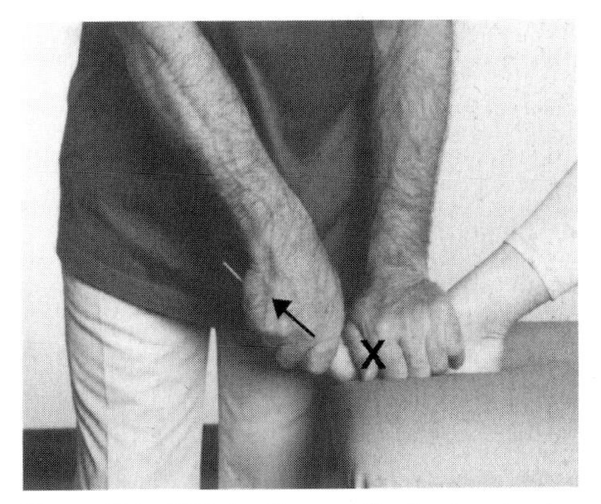

Figure 43d – MCP traction-mobilization in extension

■ Figure 43c: Flexion progression

Objective

- To increase flexion range-of-motion in a DIP, PIP, or MTP joint.

Starting position

- The plantar side of the patient's foot rests on a wedge.
- Position the targeted DIP, PIP, or MTP joint close to its end range-of-motion in flexion.

Hand placement and fixation

- **Therapist's stable hand (left):** Fixate the patient's foot against the wedge with your hand; grip with your thenar eminence just proximal to the targeted joint space.
- **Therapist's moving hand (right):** Hold the patient's toe in your hand; grip with your fingers just distal to the targeted joint space.

Procedure

- Apply a Grade III distal traction movement to the phalanx.

■ Figure 43d: Extension progression for the MTP joints

- The patient's foot rests on a wedge with the MTP joint positioned close to its end range-of-motion in extension.
- Apply a Grade III distal traction movement to the phalanx.

Toe plantar glide
for restricted flexion

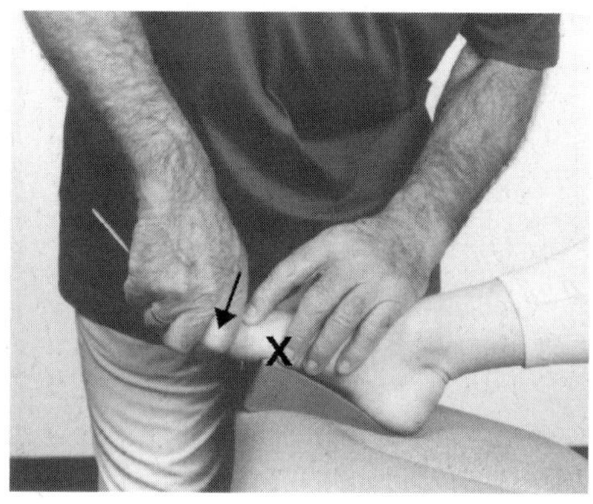

Figure 44a – test and mobilization in resting position

■ Figure 44a: Test and mobilization in resting position

Objective

- To evaluate the quantity and quality of plantar glide joint play in a DIP, PIP, or MTP joint, including end-feel.
- To increase DIP, PIP, or MTP joint flexion (Concave Rule).

Starting position

- The plantar side of the patient's foot rests on a wedge.
- Position the targeted DIP, PIP, or MTP joint in its resting position.

Hand placement and fixation

- **Therapist's stable hand (left):** Fixate the patient's foot against the wedge with your hand; grip the patient's mid-foot with your fingers just proximal to the targeted joint space.
- **Therapist's moving hand (right):** Hold the patient's toe in your hand; grip with your fingers just distal to the targeted joint space.

Procedure

- Apply a Grade II or III palmar glide movement to the phalanx.
- Palpate the joint space with your index finger during the movement.

Toe plantar glide

for restricted flexion (cont'd)

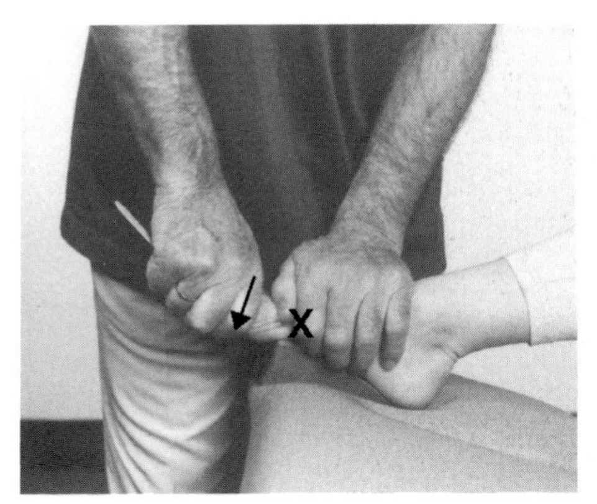

Figure 44b – mobilization in resting position

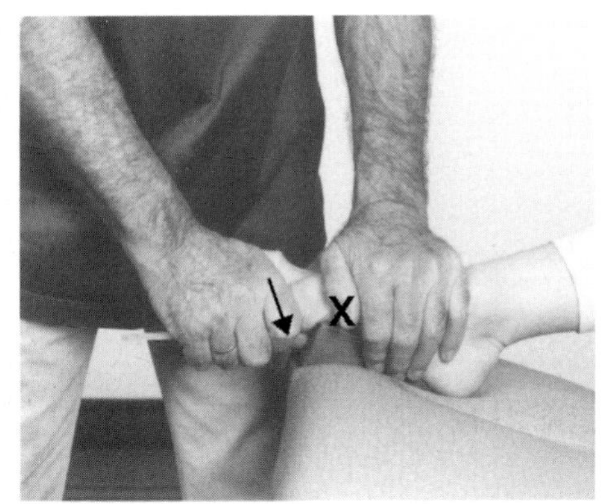

Figure 44c – mobilization in flexion

■ Figure 44b: Mobilization in resting position

Objective

- To increase flexion range-of-motion in a DIP, PIP, or MTP joint (Concave Rule).

Starting position

- The plantar side of the patient's foot rests on a wedge.
- Position the targeted DIP, PIP, or MTP joint in its resting position.

Hand placement and fixation

- **Therapist's stable hand (left):** Fixate the patient's foot against the wedge with your hand; grip with your thenar eminence just proximal to the targeted joint space.
- **Therapist's moving hand (right):** Hold the patient's toe in your hand; grip with your fingers just distal to the targeted joint space.

Procedure

- Apply a Grade III plantar glide movement to the phalanx.

■ Figure 44c: Flexion progression

- Apply a Grade III plantar glide movement with the targeted toe joint positioned close to its end range-of-motion in flexion.

Toe dorsal glide
for restricted extension

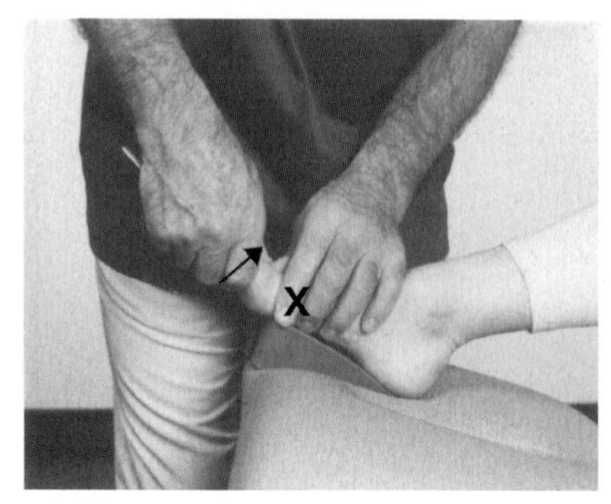

Figure 45a – test and mobilization in resting position

■ Figure 45a: Test and mobilization in resting position

Objective

- To evaluate the quantity and quality of dorsal glide joint play in a DIP, PIP, or MTP joint, including end-feel.
- To increase DIP, PIP, or MTP extension (Concave Rule).

Starting position

- The plantar side of the patient's foot rests on a wedge.
- Position the targeted DIP, PIP, or MTP joint in its resting position.

Hand placement and fixation

- **Therapist's stable hand (left):** Fixate the patient's foot against the wedge with your hand; grip the patient's mid-foot with your fingers just proximal to the targeted joint space.
- **Therapist's moving hand (right):** Hold the patient's toe in your hand; grip with your fingers just distal to the targeted joint space.

Procedure

- Apply a Grade II or III dorsal glide movement to the phalanx.
- Palpate the joint space with your index finger during the movement.

Toe dorsal glide
for restricted extension (cont'd)

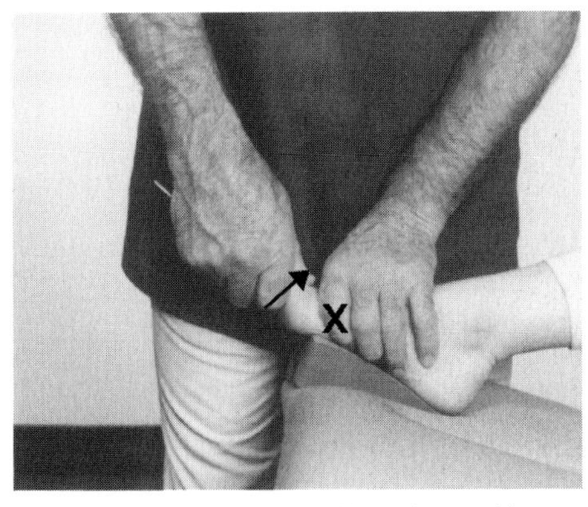

Figure 45b – mobilization in resting position

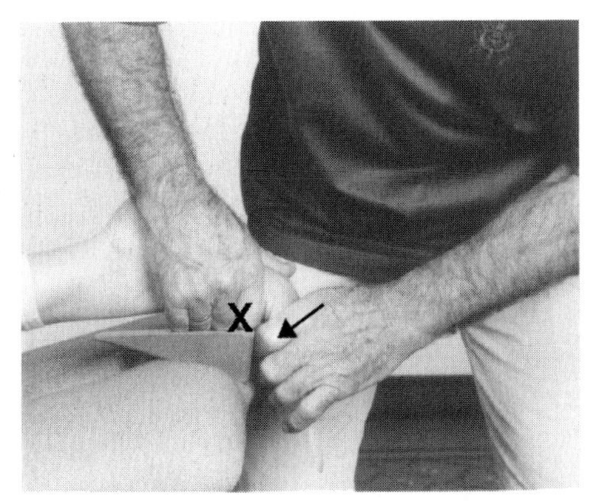

Figure 45c – mobilization in extension

■ Figure 45b: Mobilization in resting position

Objective

- To increase extension range-of-motion in a DIP, PIP, or MTP joint (Concave Rule).

Starting position

- The plantar side of the patient's foot rests on a wedge.
- Position the targeted DIP, PIP, or MTP joint in its resting position.

Hand placement and fixation

- **Therapist's stable hand (left):** Fixate the patient's proximal joint partner against the wedge with your hand; apply pressure with your thenar eminence just proximal to the targeted joint space.
- **Therapist's moving hand (right):** Hold the patient's toe in your hand; grip with your fingers just distal to the targeted joint space.

Procedure

- Apply a Grade III dorsal glide movement to the phalanx.

■ Figure 45c: Extension progression

- Apply a Grade III dorsal glide movement to the phalanx with the targeted toe joint positioned close to its end range-of-motion in extension. The dorsal side of the patient's foot rests on the wedge.

Toe tibial glide
for restricted abduction, flexion & extension

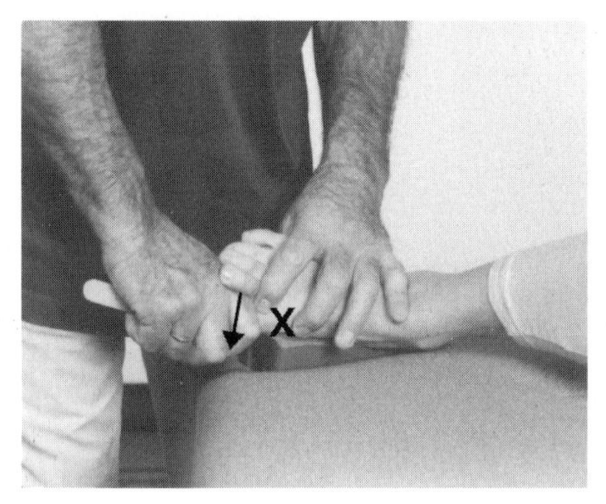

Figure 46a – test and mobilization in resting position

■ Figure 46a: Test and mobilization in resting position

Objective

- To evaluate the quantity and quality of tibial glide joint play in a DIP, PIP, or MTP joint, including end-feel.
- To increase toe flexion, extension, or tibial flexion (abduction) in a MTP joint.

Starting position

- The tibial side of the patient's foot rests on a wedge.
- Position the targeted DIP, PIP, or MTP joint in its resting position.

Hand placement and fixation

- **Therapist's stable hand (left):** Fixate the patient's forefoot against the wedge with your hand; grip with your fingers just proximal to the targeted joint space.
- **Therapist's moving hand (right):** Hold the patient's toe in your hand; grip with your fingers just distal to the targeted joint space; improve your "handle" on the toe by gripping with tape, a paper towel, or a tongue depressor taped to the toe.

Procedure

- Apply a Grade II or III tibial glide movement to the phalanx.
- Palpate the joint space with your index finger during the movement.

■ Toe fibular glide: test for restricted flexion and extension (not shown)

- Follow the same procedure for testing fibular glide joint play. Apply a Grade II or III fibular glide movement.

Toe tibial glide

for restricted abduction, flexion & extension (cont'd)

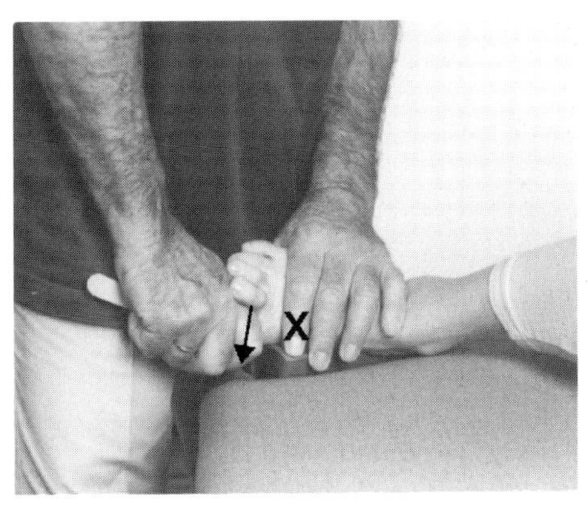

Figure 46b – mobilization in resting position

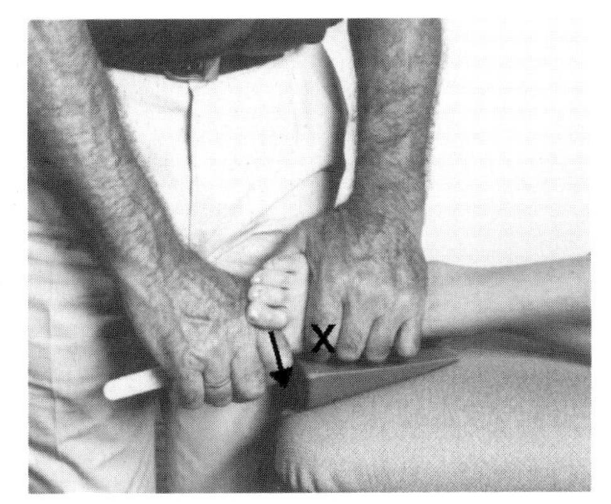

Figure 46c - mobilization for abduction

■ Figure 46b: Mobilization in resting position

Objective

- To increase tibial flexion (abduction) range-of-motion in a MTP joint (Concave Rule).
- To increase flexion or extension range-of-motion in a DIP, PIP, or MTP joint

Starting position

- The tibial side of the patient's foot rests on a wedge.
- Position the targeted DIP, PIP, or MTP joint in its resting position.

Hand placement and fixation

- **Therapist's stable hand (left):** Fixate the patient's forefoot against the wedge with your hand; grip with your fingers just proximal to the targeted joint space.
- **Therapist's moving hand (right):** Hold the patient's toe in your hand; grip with your fingers just distal to the targeted joint space; improve your "handle" on the toe by gripping with tape, a paper towel, or a tongue depressor taped to the toe.

Procedure

- Apply a Grade III tibial glide movement to the phalanx.

■ Figure 46c: Abduction progression

- Apply a Grade III tibial glide movement with the MTP joint positioned near its end range-of-motion into abduction. In cases of extreme hypomobility, the MTP joint may remain in an adducted position.

■ Toe fibular glide: mobilization and progression (not shown)

- Follow the same procedure for fibular glide. Apply a Grade III fibular glide movement; may be performed in the resting position and progressed to end range-of-motion positioning into flexion or extension.

■ Notes

CHAPTER 16

METATARSALS

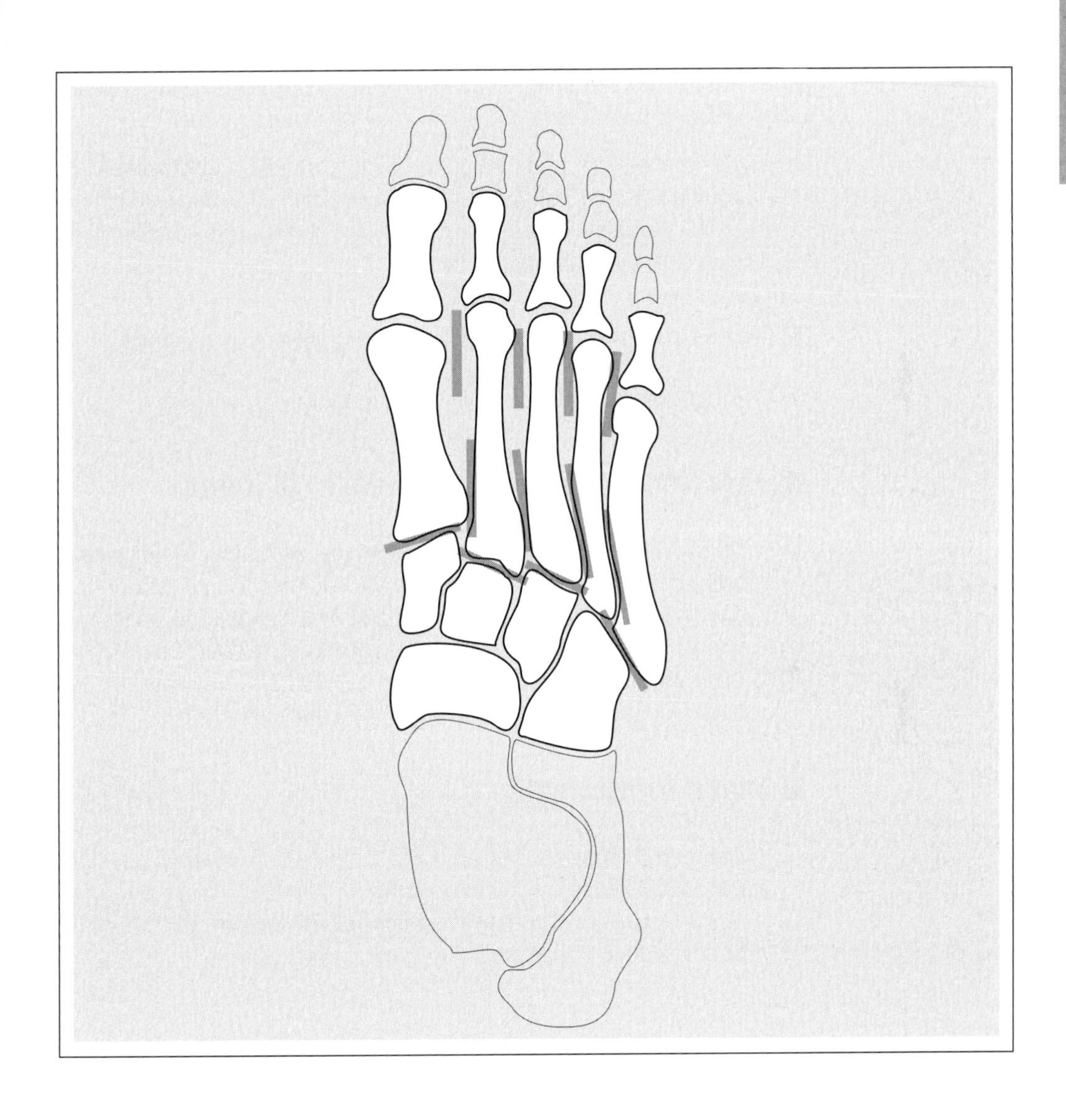

16 Metatarsals

(metatarsus)

■ Functional anatomy and movement

The metatarsus consists of five metatarsal bones (metatarsalia, abbreviated MT), one for each toe. Each metatarsal bone has a head (caput) or distal end with a convex surface, a body (corpus) and a base (basis) or proximal end with a concave surface.

The metatarsal joints are anatomically simple and mechanically compound plane gliding joints (amphiarthroses, modified sellar). Small gliding movements, synchronized with movements of the foot, take place in the metatarsus.

■ Proximal metatarsals (intermetatarsal joints)

Plane joints lie between the bases of the MT bones I-V.

■ Tarsometatarsals (tarsometatarsal joints)

Joints which are almost plane lie between the base of the metatarsal bones (functionally concave) and the adjacent row of tarsal bones (functionally convex). Figure 49 illustrates how the MT bones I, II, and III articulate with the three cuneiform bones (medial, intermediate and lateral, abbreviated C1, C2, and C3) and that the MT bones IV and V together articulate with the cuboid.

■ Distal metatarsals

There are no joints between the heads of the metatarsals. The metatarsal heads are joined together by the deep transverse metatarsal ligaments. Their movements follow the movements of the proximal metatarsals, but with greater range.

Metatarsals

Bony palpation
- Metatarsal bones I-V
- Distal row of tarsals (cuneiform I-III, cuboid)
- Joint spaces of tarsometatarsal joints I-V

Ligaments
- Metatarsal ligaments (dorsal, interosseous and plantar)
- Tarsometatarsal ligaments (plantar and dorsal)
- Interosseous cuneometatarsal ligaments

Bone movement and axes
Intermetatarsal joints:
- There are no defined axes for the small movements that occur in these joints. Intermetatarsal joint movements increase and decrease the curve of the transverse metatarsal arch.

  As the curve of the transverse metatarsal arch increases, the tarsals glide in a plantar direction with relation to metatarsal II.

  As the curve of the transverse metatarsal arch decreases, the tarsals glide in a dorsal direction with relation to metatarsal II.

Tarsometatarsal joints:
- Plantar - dorsal flexion: around a tibiofibular axis through the distal part of the cuneiform I-III and the cuboid.

End feel
- Firm

Joint movement (gliding)
- Concave rule

Treatment plane
- *Distal and proximal intermetatarsal I-V:* parallel between the metatarsals
- *Tarsometatarsal:* on the concave joint surface at the base of the metatarsal

Zero position:
- Unknown

Resting position
- Unknown

Close-packed position
- Unknown

Capsular pattern
- Unknown

Metatarsals

Metatarsal examination scheme

(Refer to Chapters 3 and 4 for more information on examination)

Tests of function

1. Active and passive movements, including stability tests and end-feel

Tarsometatarsal joints I-V

Flexion - extension	very slight movement

2. Translatoric joint play movements, including end-feel

Tarsometatarsal joint I-V

Traction - compression	(adapt from Metacarpals, Figures 6a, 7a)

Distal intermetatarsal

Gliding plantar - dorsal	(Figure 47a)

Proximal intermetatarsal

Gliding plantar - dorsal	(Figure 48a)

Trial treatment

Tarsometatarsal traction (adapt from *Metacarpals*, Figures 6b, 7b)

Metatarsal techniques

Recommended mobilization sequence for the metatarsals

1. Proximal intermetatarsal plantar glide (Figure 48)
2. Distal intermetatarsal glide (Figure 47)

Distal intermetatarsal plantar glide

for hypomobility

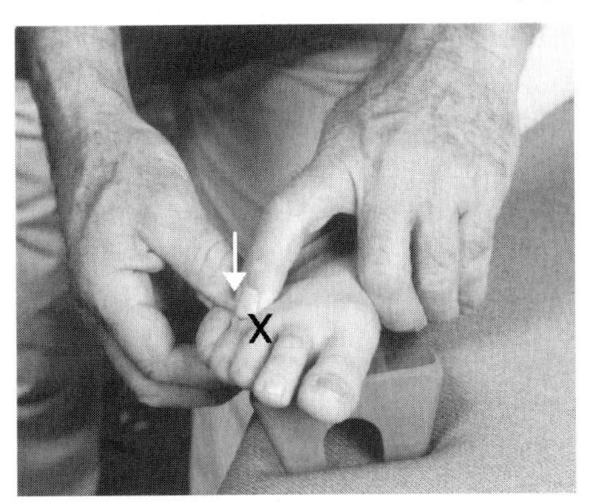

Figure 47a – test

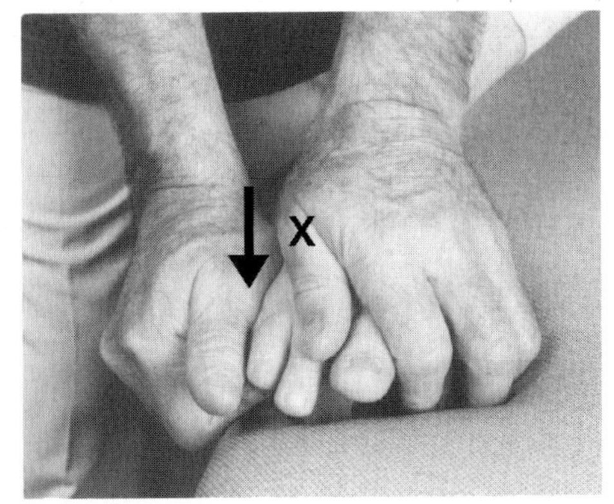

Figure 47b – mobilization

■ Figure 47a: Test

Objective

- To evaluate the quantity and quality of metatarsal plantar glide joint play, including end-feel.

Starting position

- The plantar side of the patient's foot rests on the treatment surface or a wedge, with the targeted metatarsal extending over the lateral edge.

Hand placement and fixation:

- **Fixation:** The medial metatarsal is supported on a wedge (MT III shown) or with the therapist's hand.
- **Therapist's stable hand (left):** Hold the patient's foot from the medial side; place your palpating finger in the targeted syndesmosis (MT III-IV shown).
- **Therapist's moving hand (right):** Grip around the patient's adjacent metatarsal (MT IV shown).

Procedure

- Apply a Grade II or III plantar glide movement to the adjacent metatarsal (MT IV shown).

■ Figure 47b: Mobilization

- Fixate metatarsal III with your thenar eminence on the dorsal surface of the foot; apply a Grade III plantar glide movement with your right thenar eminence pressing down on the adjacent metatarsal (MT IV/V shown) to increase metatarsal mobility and stretch the syndesmosis.

■ Alternate mobilization technique (not shown)

- Use the same grip; apply a Grade III dorsal glide movement.

Metatarsals

Proximal intermetatarsal plantar glide
for hypomobility

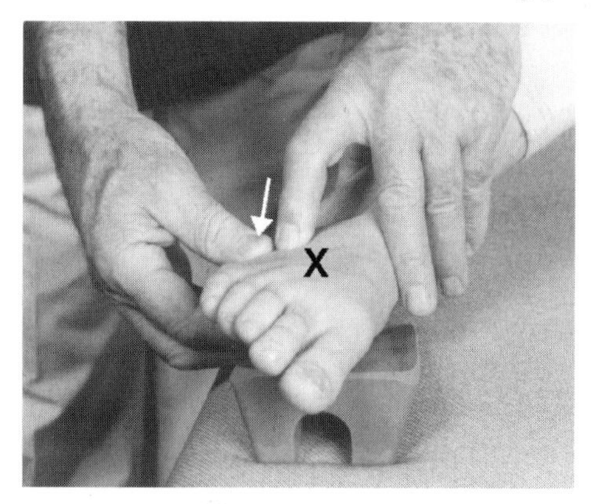

Figure 48a – test

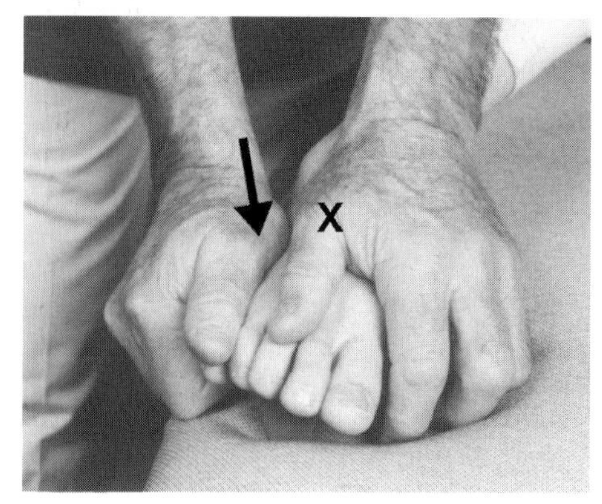

Figure 48b – mobilization

■ Figure 48a: Test

Objective

- To evaluate the quantity and quality of metatarsal plantar glide joint play, including end-feel.

Starting position

- The plantar side of the patient's foot rests on the treatment surface or a wedge, with the targeted metatarsal extending over the lateral edge.

Hand placement and fixation:

- **Fixation:** The medial metatarsal is supported on a wedge (metatarsal IV shown) or with the therapist's hand.
- **Therapist's stable hand (left):** Hold the patient's foot from the medial side; place your palpating finger in the targeted joint space (metatarsal IV-V shown).
- **Therapist's moving hand (right):** Grip around the patient's proximal metatarsal (metatarsal V shown).

Procedure

- Apply a Grade II or III plantar glide movement to the adjacent metatarsal (metatarsal V shown).

■ Figure 48b: Mobilization

- Fixate metatarsal III with your thenar eminence on the dorsal surface of the foot; apply a Grade III plantar glide movement with your right thenar eminence pressing down on the base of the adjacent metatarsal (metatarsal IV/V shown) to increase forefoot mobility by increasing intermetatarsal plantar glide.

■ Alternate mobilization technique (not shown)

- Use the same grip; apply a Grade III dorsal glide movement.

CHAPTER 17

FOOT & ANKLE

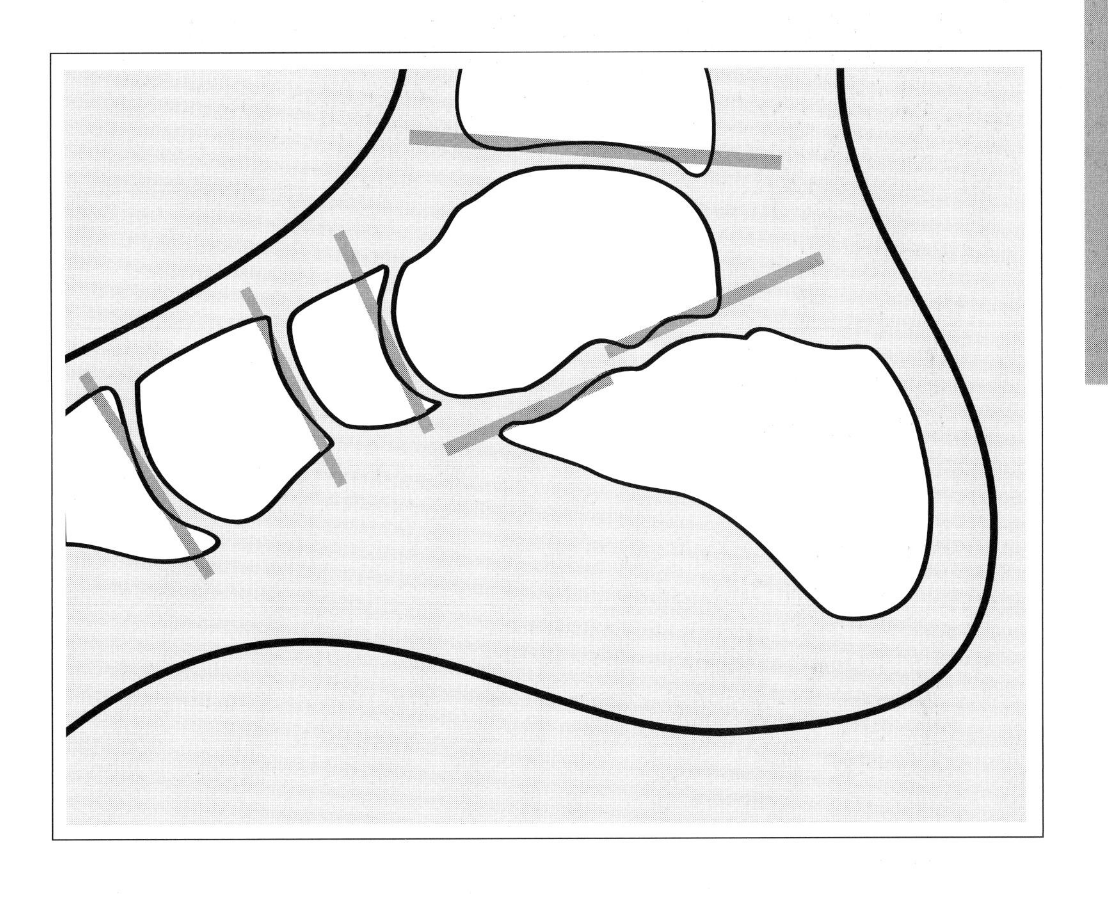

17 Foot and ankle

Functional anatomy and movement

The foot (tarsus) consists of seven tarsal bones (see figure 49). The ankle joint (art. talocruralis) includes the talus articulating with the distal aspect of the tibia and fibula. The bones of the foot are: three cuneiform bones (ossa cuneiformia I, II and III); cuboid (os cuboideum); navicular (scaphoid = os naviculare); talus (astragalus); and calcaneus.

Tarsal joints

For testing and treatment purposes, we divide the tarsal joints into the following functional units:

Cuneonavicular joint

The cuneonavicular joint consists of three convex facets on the navicular and three concave facets on the cuneiform bones.

Cuboid-3rd cuneiform/navicular

Medially, the slightly concave cuboid articulates with the convex 3rd cuneiform and navicular.

Calcaneocuboid joint

The calcaneocuboid joint has a saddle-shaped joint surface. The concave surface on the cuboid guides movement for flexion and extension. The convex surface on the cuboid guides movement for abduction and adduction.

Talonavicular joint

The convex anterior surface of the talus articulates with the corresponding concave surface of the navicular.

Talocalcaneal joint, anterior and middle

The convex inferior/anterior surfaces of the talus articulate with the corresponding concave surfaces on the calcaneus.

Subtalar joint (talocalcaneal joint, posterior)

The subtalar joint is an anatomically simple and mechanically compound joint. The concave surface on the inferior/posterior surface of the talus articulates with a corresponding convex surface on the superior calcaneus.

■ Talocrural joint

The talocrural joint, between the talus and the distal surfaces of the tibia and fibula, is an anatomically and mechanically simple uniaxial joint (ginglymus, modified sellar). The trochlear surface of the talus is broader anteriorly so that during foot dorsal flexion the talus pushes the ankle mortise apart. In this position the talus fits tightly in the recess created by the tibia, fibula and the tibiofibular syndesmosis, which restricts movement of the talus and makes the joint more stable.

Foot

Bony palpation

- Cuneiform I, II and III
- Navicular
- Cuboid
- Calcaneus
- Talus
- Joint spaces between the tarsal bones
- Navicular tuberosity
- Sustentaculum tali
- Medial malleolus
- Lateral malleolus

Ligaments

- Plantar and dorsal cuneonavicular ligaments
- Plantar and dorsal intercuneiform ligaments
- Interosseous tarsal ligaments
- Cuboideonavicular (plantar and dorsal) ligaments
- Cuneocuboid (plantar and dorsal) ligaments
- Calcaneocuboid ligament (lateral half of the bifurcate ligament)
- Long plantar ligament
- Calcaneonavicular ligament (medial half of the bifurcate ligament)
- Talocalcaneal ligaments (lateral, medial, and interosseous ligaments)
- Deltoid ligament (medially)
- Calcaneofibular ligament (laterally)
- Talofibular ligament (anterior and posterior)

Bone movements and axes (see Figure 49)

- Plantar-dorsal flexion occurs primarily at the talocrural joint around a tibiofibular axis through the convex joint partner (talus).
- Pronation-supination occurs primarily in the forefoot around a longitudinal axis through metatarsal II. This movement is much greater when performed passively.
- Inversion-eversion occurs primarily between the talus and calcaneus and the talus and navicular around an oblique axis through the calcaneus and talus. Inversion is the combined movement of supination-adduction-plantar flexion; eversion is the combined movement of pronation-abduction-dorsal flexion.

Joint movement (gliding)

- Apply the Concave Rule or Convex Rule according to whether the mobilization technique moves the concave or convex surface of the targeted bone.

Treatment plane

- Lies on the concave surface of the targeted joint.

Zero position:

- The fibular side of the foot forms a right angle with the longitudinal axis through the leg.
- A line from the anterior superior iliac spine through the patella passes through the second toe.

Resting position:

- Approximately 10° plantar flexion and midway between maximal inversion and eversion.

Close-packed position:

- Metatarsus and tarsus: maximal inversion
- Talocrural joint: maximal dorsal flexion

Capsular pattern:

- Plantar flexion - dorsal flexion

Foot examination scheme

(Refer to Chapters 3 and 4 for more information on examination)

Tests of function

1. Active and passive movements, including stability tests and end-feel

Tarsal joints

Flexion	
Extension	
Pronation	10°
Supination	20°

Subtalar joints

Inversion	40°
Eversion	20°

2. Translatoric joint play movements, including end-feel

Tarsal joints

Traction - compression		(Figure 50a, 50b)
Gliding	Plantar	(Figure 50a, 51a, 52a, 53a)
	Dorsal	(Figure 50c, 50d)

Subtalar joints

Traction - compression		(Figure 54a)
Gliding	Distal	(Figure 54c)
	Tibial	(Figure 55a)
	Fibular	(Figure 55c)

3. Resisted movements

Eversion
Peronei

Inversion
Tibialis posterior
Triceps surae

4. Passive soft tissue movements

Physiological
Accessory

5. Additional tests

Trial treatment

Tarsal joint glide	(Figure 49, 50a, 50b, 50c, 50d)
Subtalar joint	(Figure 54b)

Ankle examination scheme

Tests of function

Talocrural joint

1. **Active and passive movements, including stability and end-feel**

Plantar flexion	40°
Dorsal flexion	20°

2. **Translatoric joint play movements, including end-feel**

Traction - compression	(Figure 56a)
Gliding	
Anterior	(Figure 57a)
Posterior	(Figure 59a)

3. **Resisted movements**

   Plantar flexion

   Triceps surae

   Flexor hallucis longus

   Flexor digitorum longus

   Dorsal flexion

   Tibialis anterior

   Extensor hallucis longus

   Extensor digitorum longus

4. **Passive soft tissue movements**

   Physiological

   Accessory

5. **Additional tests**

Trial treatment

Talocrural joint traction	(Figure 56b)

Foot and ankle glide tests

Recommended sequence

■ **Figure 49: Recommended glide test sequence for the foot and ankle**

Movements in the middle of the foot (distally)

Fixate the third and second cuneiforms and move:

1. Metatarsal III
2. Metatarsal II

Movements on the medial side of the foot (Figure 50a)

Fixate the first cuneiform and move:

3. Metatarsal I

Fixate navicular and move:

4. The first, second, and third cuneiforms

Fixate talus and move:

5. Navicular

Movements on the lateral side of the foot around the cuboid

Fixate cuboid and move:

6. Metatarsals IV and V (Figure 51a)

Fixate the navicular and the third cuneiform from the medial side and move:

7. Cuboid (Figure 52a)

Fixate calcaneus and move:

8. Cuboid (Figure 53a)

Movements between the talus and calcaneus

Fixate talus and move:

9. Calcaneus (Figure 55a)

Movements in the ankle joint

Fixate the leg and move:

10. Talus, or (Figure 56a)
11. Fixate the talus and move the leg (Figure 57a)

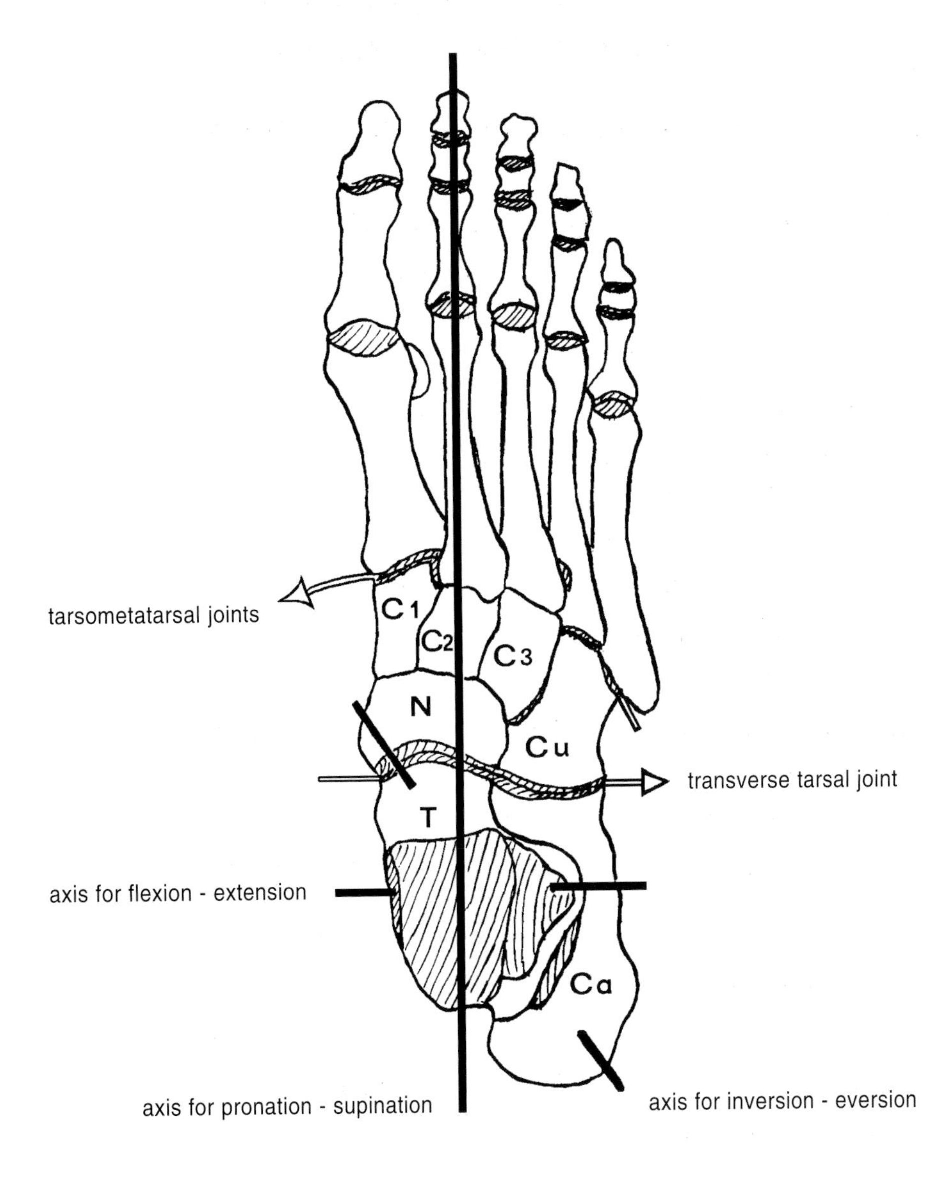

Figure 49
Tarsal bones and axes
(dorsal aspect of the right foot)

C1	=	*cuneiform I*
C2	=	*cuneiform II*
C3	=	*cuneiform III*
Cu	=	*cuboid*
N	=	*navicular*
T	=	*talus*
Ca	=	*calcaneus*

■ Foot and ankle techniques

Foot

Cuneonavicular plantar glide

for restricted plantar flexion

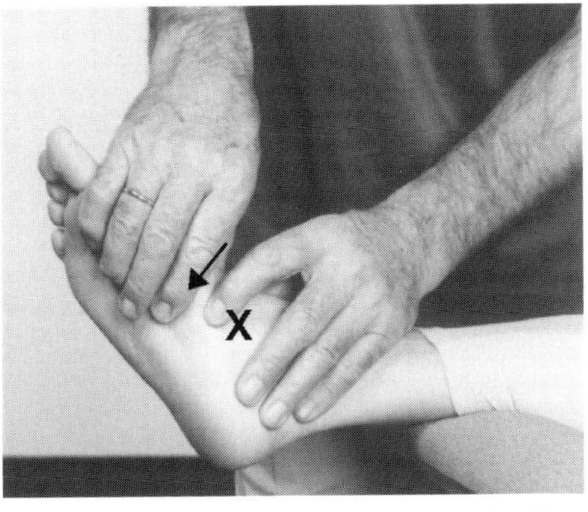

Figure 50a – test and mobilization in resting position

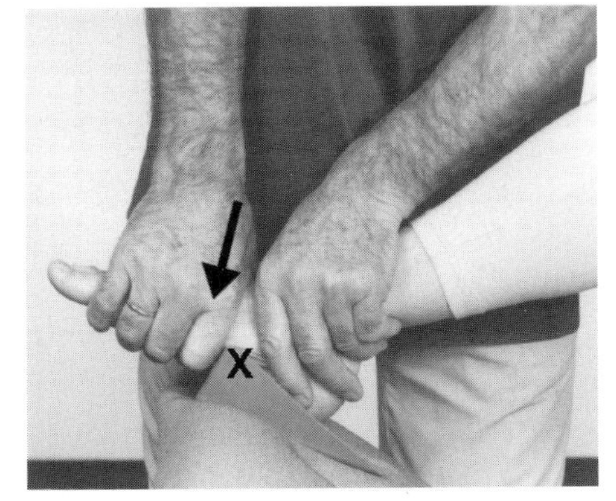

Figure 50b – mobilization in resting position

Foot

■ Figure 50a: Test and mobilization in resting position

Objective

- To evaluate the quantity and quality of plantar glide joint play of the cuneiform in relation to the navicular, including end-feel.
- To increase foot plantar flexion range-of-motion (Concave Rule).

Starting position

- The posterior side of the patient's legs rests on the treatment surface.
- Position the joint in its resting position.

Hand placement and fixation

- **Therapist's stable hand (left):** Hold the patient's foot; grip the patient's navicular bone with your fingers; fixate the patient's foot against treatment surface.
- **Therapist's moving hand (right):** Hold the patient's forefoot; grip the cuneiform I with your fingers just distal to the joint space.

Procedure

- Apply a Grade II or III plantar glide movement to the cuneiform bone; palpate the joint space with your left index finger.

Notes

- Use the same procedure to mobilize the MT I - cuneiform I and the talonavicular joint.
- All joints on the medial side of the foot can also be tested with traction and compression, and treated with traction.

■ Figure 50b: Mobilization in resting position

- Rest the plantar side of the patient's foot, including the navicular, on the wedge.
- Hold the patient's foot with your hand; grip with your index finger and second metacarpal over the first cuneiform
- Apply a Grade III plantar glide movement by leaning with your body over your extended arm.

Navicular-talus dorsal glide

for restricted dorsal flexion

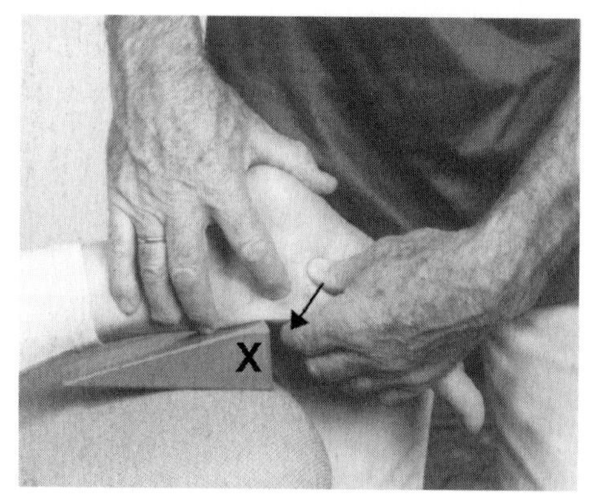

Figure 50c – test and mobilization in resting position

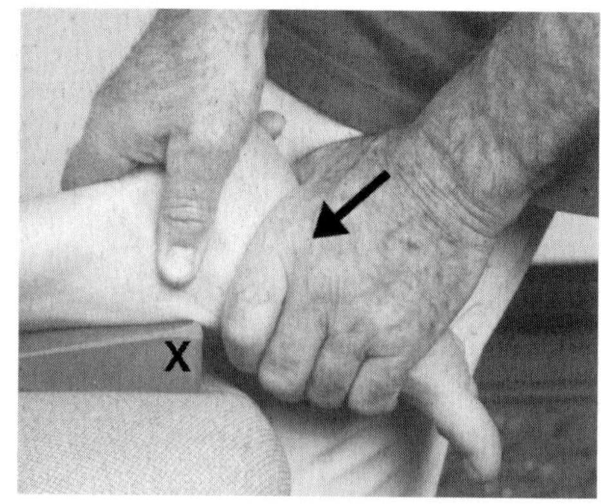

Figure 50d – mobilization in resting position

■ Figure 50c: Test and mobilization in resting position

Objective

- To evaluate the quantity and quality of dorsal glide joint play of the navicular in relation to the talus, including end-feel.
- To increase foot dorsal flexion range-of-motion (Concave Rule).

Starting position

- The anterior side of the patient's leg rests on the treatment wedge.
- Position the joint in its resting position.

Hand placement and fixation

- **Therapist's stable hand (right):** Hold the patient's distal leg against the wedge; place your palpating finger in the joint space.
- **Therapist's moving hand (left):** Hold the patient's mid-foot in your hand; grip with your fingers surrounding the navicular bone.

Procedure

- Apply a Grade II or III dorsal glide movement to the navicular bone.

Notes

- Use the same procedure to mobilize the cuneiform - MT I and cuneionavicular joint.

■ Figure 50d: Mobilization in resting position

- Hold the patient's foot with your left hand; grip with your index finger and second metacarpal over the navicular bone.
- Apply a Grade III dorsal glide movement by leaning with your body through your arm.

Metatarsals IV/V-cuboid plantar glide

for restricted plantar flexion

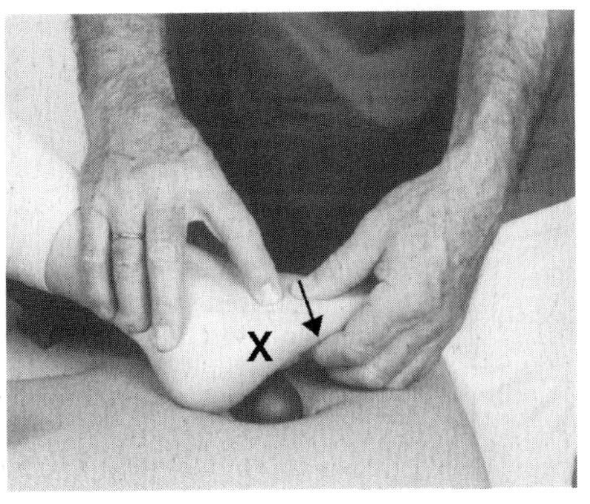

Figure 51a – test in resting position

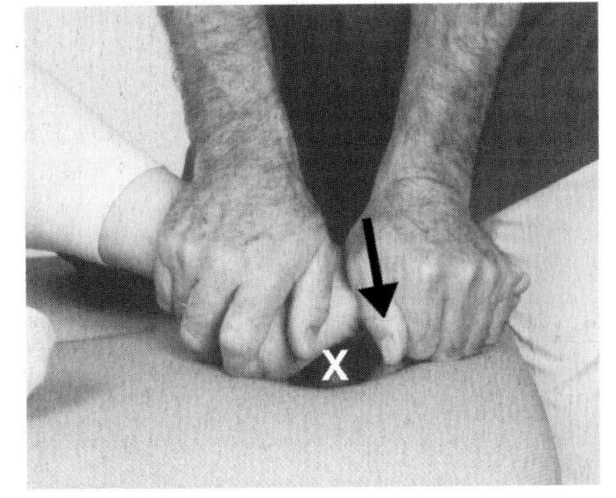

Figure 51b – mobilization in resting position

Foot

■ Figure 51a: Test in resting position

Objective

- To evaluate the quantity and quality of plantar glide joint play of metatarsals IV and V in relation to the cuboid, including end-feel.
- To increase foot plantar flexion range-of-motion (Concave Rule).

Starting position

- The tibial-plantar side of the patient's leg rests on the treatment surface.
- Position the joint in its resting position.

Hand placement and fixation

- **Fixation**: Fixate the cuboid on a wedge or sandbag.
- **Therapist's stable hand (right):** Hold the patient's ankle against the treatment surface; place your palpating finger in the joint space.
- **Therapist's moving hand (left):** Hold the patient's forefoot; grip the bases of metatarsals IV and V with your fingers.

Procedure

- Apply a Grade II or III plantar glide movement to metatarsals IV/V; palpate the joint spaces.

Notes

- All joints on the lateral side of the foot can also be tested with traction and compression, and treated with Grade I, II or III traction.

■ Figure 51b: Mobilization in resting position

- Use your right hand to fixate the cuboid bone on a wedge with metatarsals IV and V extending just past the edge.
- Grip the metatarsals with your left hand, with your thenar eminence over the bases of the metatarsals.
- Apply a Grade III plantar glide movement by leaning your body over your extended arm.

Cuboid-cuneiform III plantar glide
for hypomobility

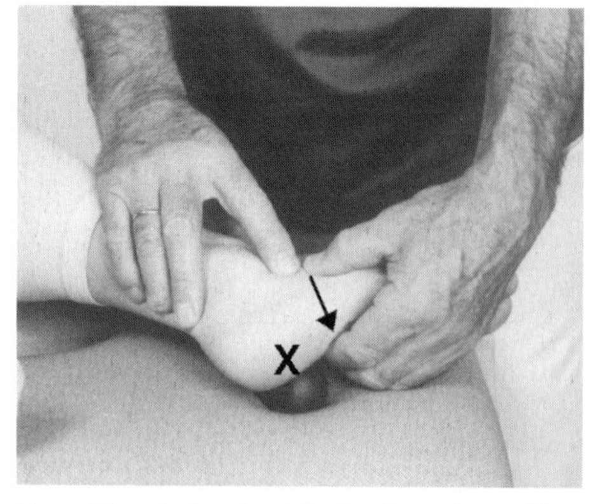

Figure 52a – test and mobilization in resting position

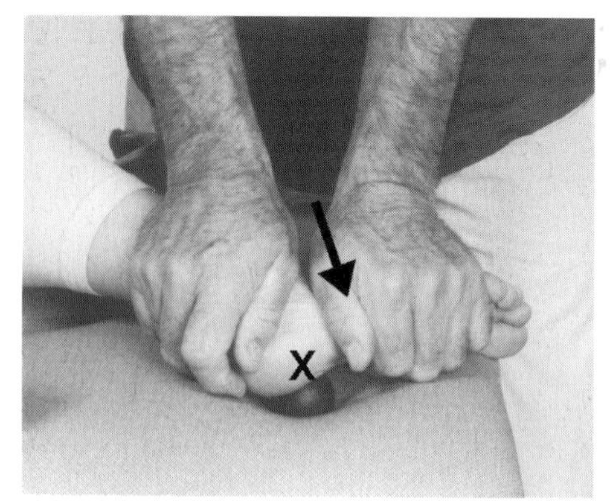

Figure 52b – mobilization in resting position

■ Figure 52a: Test and mobilization in resting position

Objective

- To evaluate the quantity and quality of plantar glide joint play of the cuboid in relation to the cuneiform III and navicular bones, including end-feel.
- To increase mid-foot mobility.

Starting position

- The tibial-plantar side of the patient's leg rests on the treatment surface.
- Position the joint in its resting position.

Hand placement and fixation

- **Fixation**: Fixate cuneiform III and navicular on a wedge or sandbag.
- **Therapist's stable hand (right):** Hold the patient's ankle against the treatment surface; place your palpating finger in the joint space.
- **Therapist's moving hand (left):** Hold the patient's forefoot; grip the cuboid with your fingers.

Procedure

- Apply a Grade II or III plantar glide movement to the cuboid; palpate the joint space.

■ Figure 52b: Mobilization in resting position

- Use your right hand to fixate the patient's ankle on a wedge with cuboid extending just past the edge.
- Grip the forefoot with your left hand, with your thenar eminence over the cuboid.
- Apply a Grade III plantar glide movement by leaning your body over your extended arm.

Cuboid-calcaneus plantar glide
for hypomobility

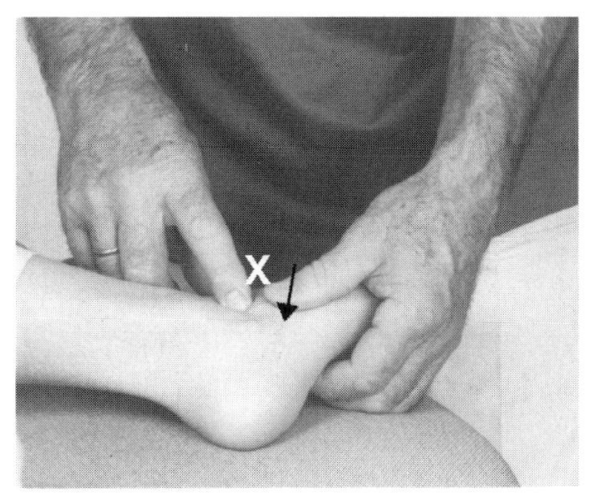

Figure 53a – test and mobilization in resting position

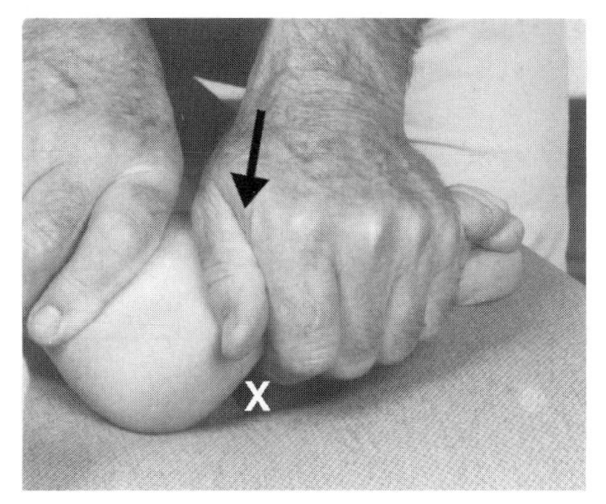

Figure 53b – mobilization in resting position

Foot

■ Figure 53a: Test and mobilization in resting position

Objective

- To evaluate the quantity and quality of plantar glide joint play of the cuboid in relation to the calcaneus, including end-feel.
- To increase mid-foot mobility.

Starting position

- The tibial-plantar side of the patient's leg rests on the treatment surface.
- Position the joint in its resting position.

Hand placement and fixation

- **Fixation**: Fixate the calcaneus on a wedge or sandbag.
- **Therapist's stable hand (right):** Hold the patient's ankle against the treatment surface; place your palpating finger in the joint space.
- **Therapist's moving hand (left):** Hold the patient's forefoot; grip the cuboid with your fingers.

Procedure

- Apply a Grade II or III plantar glide movement to the cuboid; palpate the joint space.

■ Figure 53b: Mobilization in resting position

- Use your right hand to fixate the patient's ankle on a wedge with the cuboid extending just past the edge.
- Grip the forefoot with your left hand, with your thenar eminence over the cuboid.
- Apply a Grade III plantar glide movement by leaning your body over your extended arm.

Calcaneus-talus distal traction

for pain and hypomobility

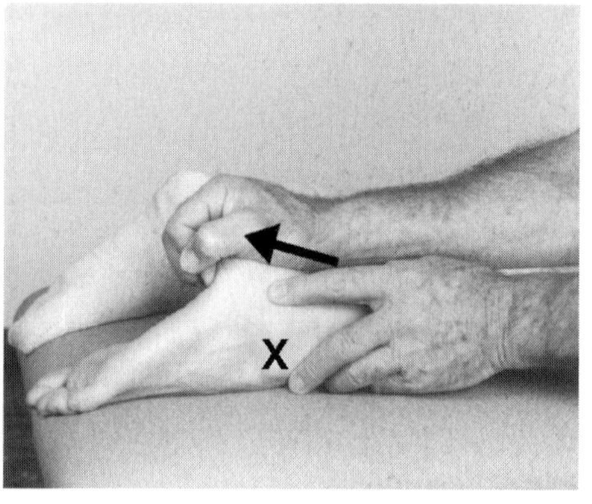

Figure 54a – test and mobilization in resting position

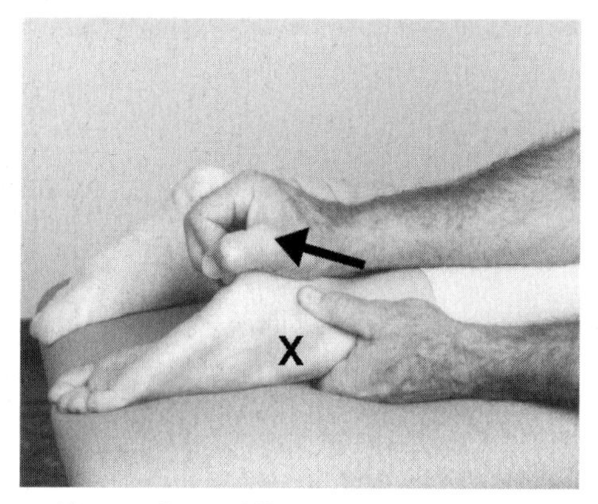

Figure 54b – mobilization in resting position

■ Figure 54a: Test and mobilization in resting position

Objective

- To evaluate the quantity and quality of distal traction joint play of the calcaneus in relation to the talus, including end-feel.
- To decrease pain or increase foot inversion or eversion range-of-motion.

Starting position

- The anterior side of the patient's leg and the dorsal side of the foot rests on the treatment surface.
- Position the subtalar joint in its resting position.

Hand placement and fixation

- **Therapist's stable hand (left):** Hold the patient's distal leg against the treatment surface; place your palpating finger in the joint space.
- **Therapist's moving hand (right):** Grip the patient's calcaneus with your thenar eminence and fingers, rest your forearm on the patient's leg.

Procedure

- Apply a Grade I, II, or III distal traction movement to the calcaneus, parallel to the long axis of the leg.

■ Figure 54b: Mobilization in resting position

- Hold the patient's distal leg with your left hand; grip around the talus from the ventral side.
- Apply a Grade III distal traction movement to the calcaneus in relation to the talus.

Calcaneus-talus distal glide

for hypomobility

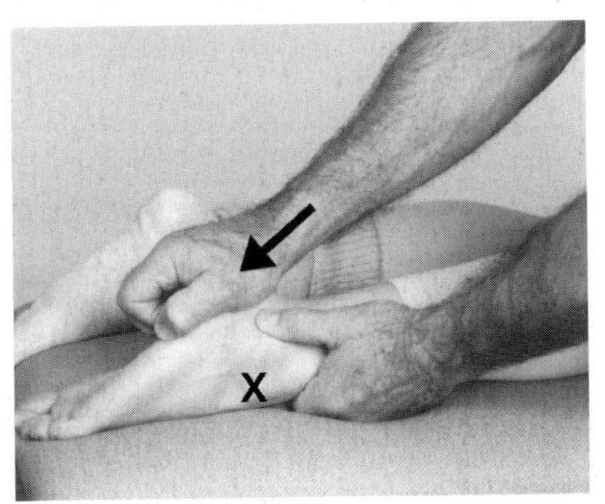

Figure 54c – test and mobilization in resting position

Foot

■ Figure 54c: Test and mobilization in resting position

Objective

- To evaluate the quantity and quality of distal glide joint play of the calcaneus in relation to the talus, including end-feel.
- To increase foot inversion and eversion.

Starting position

- The anterior side of the patient's leg and the dorsal side of the foot rests on the treatment surface.
- Position the subtalar joint in its resting position.

Hand placement and fixation

- **Therapist's stable hand (left):** Hold the patient's distal leg with your left hand; grip around the talus from the ventral side; place your palpating finger in the joint space.
- **Therapist's moving hand (right):** Grip the patient's calcaneus with your thenar eminence and fingers; position your forearm parallel to the sole of the foot.

Procedure

- Apply a Grade II or III distal glide movement to the calcaneus, parallel to the sole of the foot towards the toes.

■ Calcaneus-talus proximal glide (not shown)

- Reverse your grip to apply a Grade II or III proximal glide movement to the calcaneus away from the toes.

Calcaneus-talus tibial glide

for hypomobility

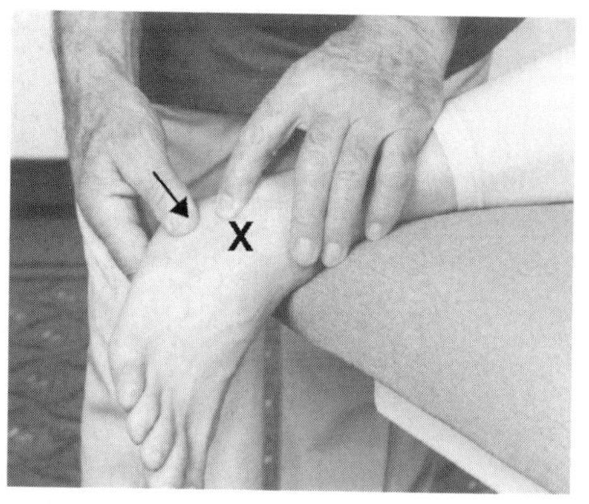

Figure 55a – test and mobilization in resting position

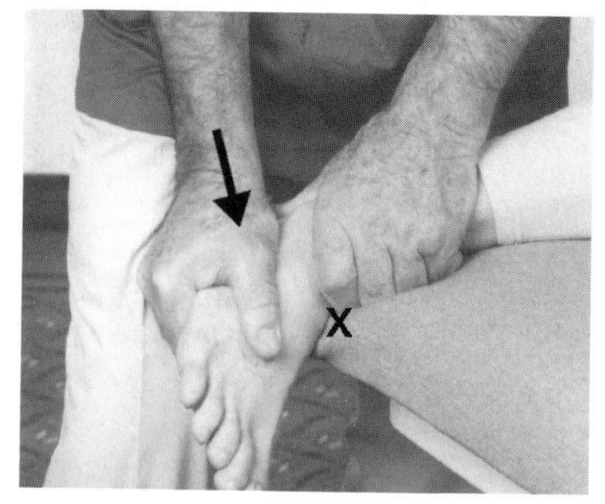

Figure 55b – mobilization in resting position

Foot

■ Figure 55a: Test and mobilization in resting position

Objective

- To evaluate the quantity and quality of tibial glide joint play in the anterior part of the talocalcaneal joint, including end-feel.
- To increase foot inversion range-of-motion (Concave Rule).

Starting position

- The tibial side of the patient's leg rests on the treatment surface.
- Position the subtalar joint in its resting position.

Hand placement and fixation

- **Therapist's stable hand (left):** Hold the patient's distal leg with your left hand; grip around the talus; place your palpating finger in the joint space.
- **Therapist's moving hand (right):** Grip the distal aspect of the patient's calcaneus with your fingers.

Procedure

- Apply a Grade II or III tibial glide movement to the calcaneus.

Note

- Use the same procedure to test tibial glide movement in the posterior part of the talocalcaneal joint. Restricted tibial glide joint play in the posterior part of the talocalcaneal joint is associated with restricted eversion (Convex Rule).

■ Figure 55b: Mobilization in resting position

- Hold the patient's distal leg with your left hand; grip around the calcaneus with your thenar eminence on the fibular side of the calcaneus.
- Apply a Grade III tibial glide movement to the calcaneus by leaning your body through your extended arm.

Calcaneus-talus fibular glide

for hypomobility

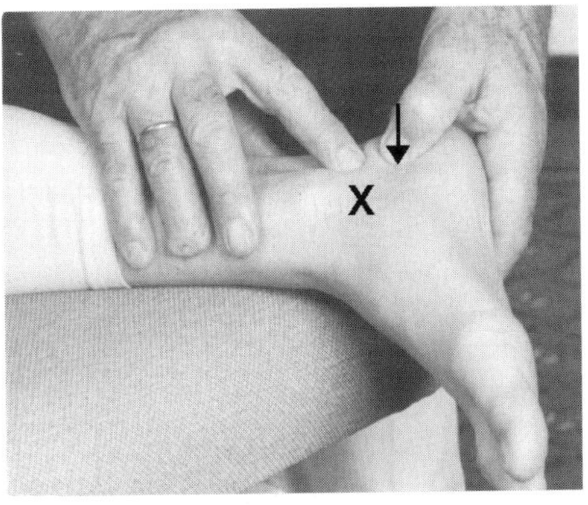

Figure 55c – test and mobilization in resting position

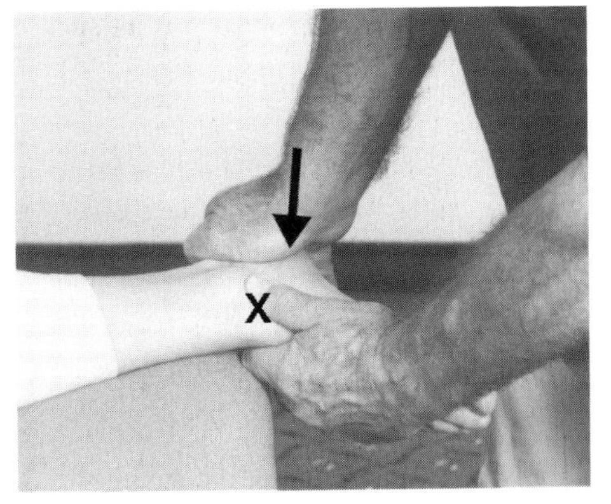

Figure 55d – mobilization in resting position

Foot

■ Figure 55c: Test and mobilization in resting position

Objective

- To evaluate the quantity and quality of fibular glide joint play in the posterior part of the talocalcaneal joint, including end-feel.
- To increase foot inversion range-of-motion (Convex Rule).

Starting position

- The fibular side of the patient's leg rests on the treatment surface.
- Position the subtalar joint in its resting position.

Hand placement and fixation

- **Therapist's stable hand (right):** Hold the patient's distal leg with your right hand; grip around the talus; place your palpating finger in the joint space.
- **Therapist's moving hand (left):** Grip the distal aspect of the patient's calcaneus with your fingers.

Procedure

- Apply a Grade II or III fibular glide movement to the calcaneus.

Note

- Use the same procedure to test fibular glide movement in the anterior part of the talocalcaneal joint. Restricted fibular glide movement in the anterior part of the talocalcaneal joint is associated with restricted eversion (Concave Rule).

■ Figure 55d: Mobilization in resting position

- Grip around the talus with your fingers.
- Apply a Grade III fibular glide movement to the calcaneus by leaning your body through your extended arm.

Talocrural traction
for pain and hypomobility

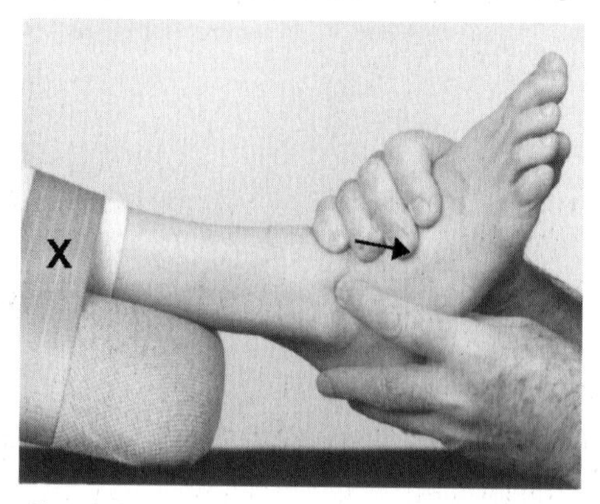

Figure 56a – test and mobilization in resting position

■ Figure 56a: Test and mobilization in resting position

Objective

- To evaluate the quantity and quality of distal traction joint play between the talus and tibia/fibula, including end-feel.
- To decrease pain or increase foot dorsal and plantar flexion range-of-motion .

Starting position

- The posterior side of the patient's leg rests on the treatment surface with the foot extended beyond the edge.
- Position the talocrural joint in its resting position.

Hand placement and fixation

- **Fixation**: Fixate the distal leg against the treatment surface with a strap.
- **Therapist's stable hand (left):** Place your palpating finger in the joint space.
- **Therapist's moving hand (right):** Grip the patient's midfoot from the tibial side with your little finger over the dorsal talus; position your forearm in line with the patient's leg.

Procedure

- Apply a Grade I, II, or III distal traction movement to the talus, parallel to the line of the leg.

Talocrural traction
for hypomobility (cont'd)

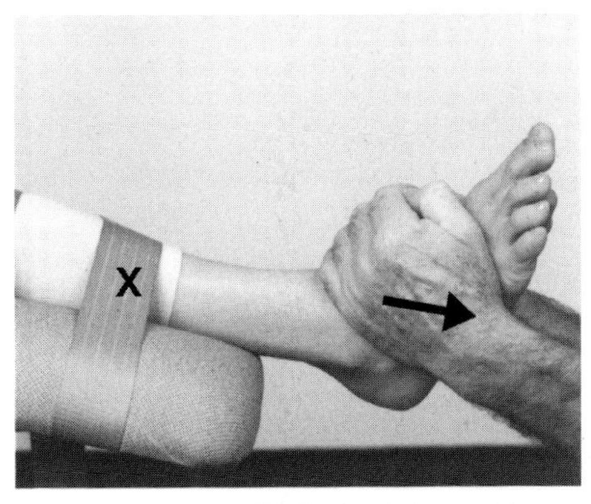

Figure 56b – mobilization in resting position

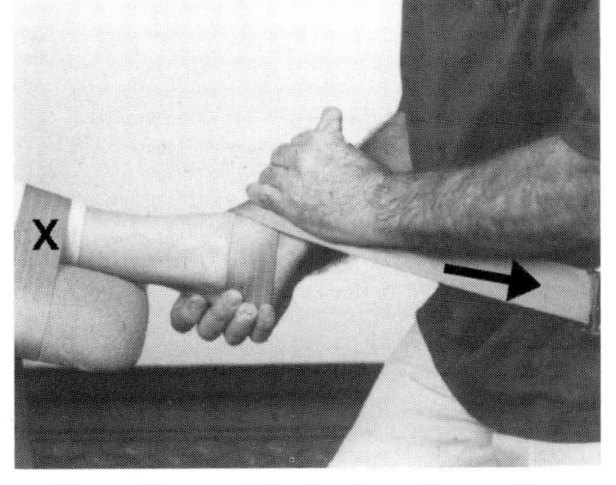

Figure 56c – mobilization in resting position

Foot

■ Figure 56b: Mobilization in resting position

Objective

- To decrease pain or increase range-of-motion between the talus and tibia/fibula.

Starting position

- The posterior side of the patient's leg rests on the treatment surface with the foot extended beyond the edge.
- Position the talocrural joint in its resting position.

Hand placement and fixation

- **Fixation**: Fixate the distal leg against the treatment surface with a strap.
- **Therapist's moving hand (right):** Grip the patient's midfoot from the tibial side with your little finger over the dorsal talus; position your forearm in line with the patient's leg; supplement your grip with your left hand.

Procedure

- Apply a Grade III distal traction movement to the talus parallel to the line of the leg, by shifting your body weight backward and pulling with both hands.

■ Figure 56c: Alternate mobilization technique in resting position

- For Grade III traction treatment of longer durations, use a mobilization strap. By attaching the mobilization strap with the "figure-8" positioned on the anterior, posterior, medial, or lateral aspect of the ankle, you can alter the effect of the mobilization.

Talocrural anterior glide
for restricted plantar flexion

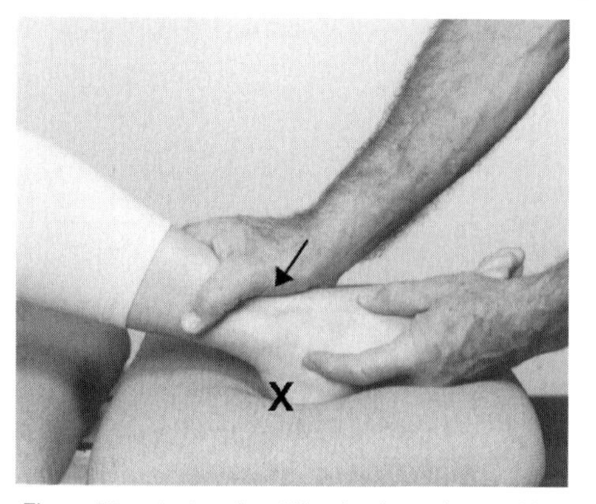

Figure 57a – test and mobilization in resting position

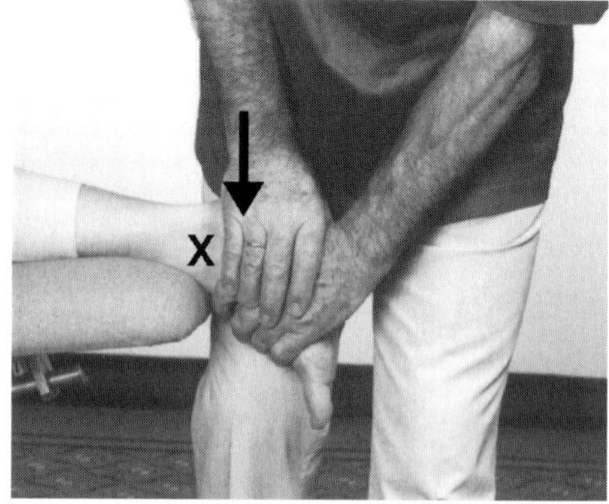

Figure 57b – mobilization in resting position

■ Figure 57a: Test and mobilization in resting position

Objective

- To evaluate the quantity and quality of anterior glide joint play of the talus in relation to the tibia/fibula.
- To increase foot plantar flexion range-of-motion (Convex Rule).

Starting position

- The patient's heel rests on the treatment surface with a bent knee.
- Position the talocrural joint in its resting position.

Hand placement and fixation

- **Fixation**: The calcaneus and, indirectly, the talus are fixated against the table.
- **Therapist's stable hand (left):** Hold the patient's foot with your hand; place your palpating finger in the joint space.
- **Therapist's moving hand (right):** Grip the distal aspect of the patient's lower leg with the heel of your hand just proximal to the joint space.

Procedure

- Apply a Grade II or III dorsal glide movement to the tibia/fibula to produce a relative anterior glide movement of the talus.

(cont'd)

Foot

■ Figure 57b Mobilization in resting position

- The patient's anterior leg rests on the treatment surface with the foot extended beyond the edge. Position the talocrural joint in its resting position.
- Grip the mid-foot from the tibial side with your left index finger around the dorsal talus; maintain a Grade I distal traction.
- Grip around the dorsal calcaneus with your right hand and supplement your grip with your left hand; position your forearm in line with the treatment plane.
- Apply a Grade III anterior glide movement to the talus by leaning your body over your extended arms and by bending your knees.

■ Plantar progression (not shown)

- Position the foot near its end range-of-motion into plantar flexion.
- Apply a Grade III anterior glide movement to the talus (Convex Rule)

Talocrural anterior glide
for restricted plantar flexion (cont'd)

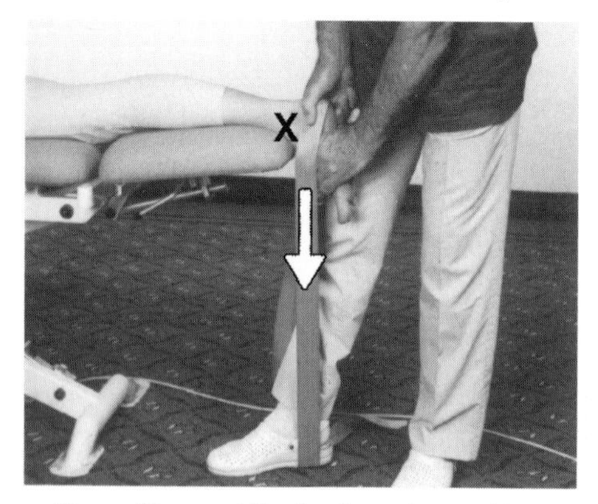

Figure 58a – mobilization in resting position

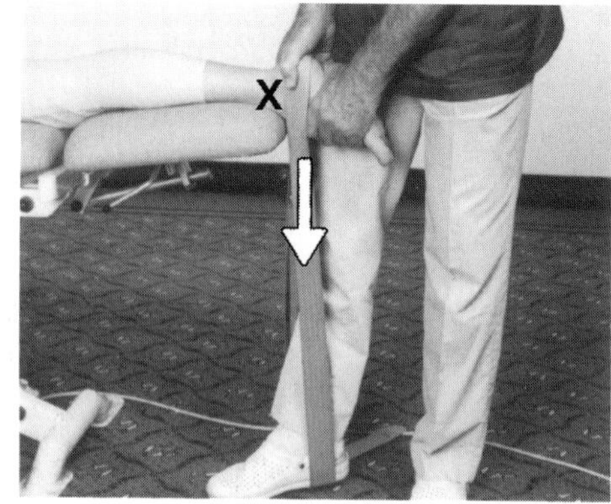

Figure 58b – mobilization in plantar flexion

Foot

■ Figure 58a: Mobilization in resting position

Objective

- To increase plantar flexion by increasing anterior glide of the talus on the tibia/fibula (Convex Rule).

Starting position

- The anterior side of the patient's leg rests on the treatment surface with the foot extended over the edge.
- Position the talocrural joint in its resting position.

Hand placement and fixation

- **Fixation**: The distal tibia and fibula are fixated against the treatment surface.
- **Therapist's moving hands:** With your left hand, grip the patient's midfoot from the tibial side with your index finger over the dorsal talus; with your right hand, grip around the posterior talus; position your right forearm in line with the treatment plane; maintain a Grade I distal traction.

Procedure

- Apply a Grade III anterior glide movement to the talus. For treatment of longer duration, use a traction strap: your heel applies the gliding force while your hands guide the movement.

■ Figure 58b: Plantar flexion progression

- Position the ankle near the end range-of-motion into plantar flexion.
- Apply a Grade III anterior glide movement to the talus (Convex Rule).

Talocrural posterior glide
for restricted dorsal flexion

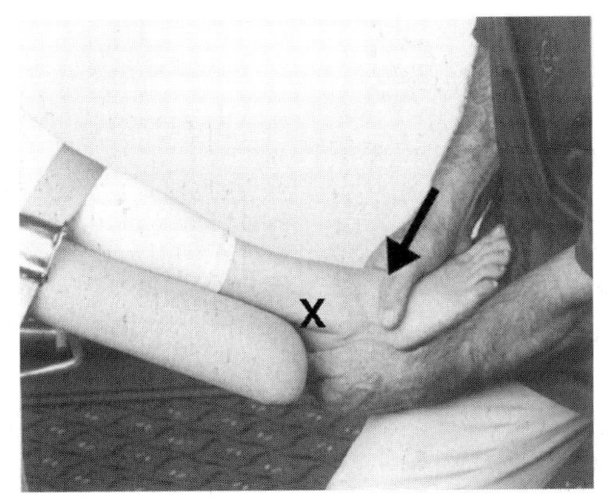

Figure 59a – Mobilization in resting position

■ Figure 59a: Mobilization in resting position

Objective

- To increase dorsal flexion by increasing dorsal glide of the talus on the tibia/fibula (Convex Rule).

Starting position

- The posterior side of the patient's leg rests on the treatment surface or wedge with the foot extended beyond the edge.
- Position the talocrural joint in its resting position.

Hand placement and fixation

- **Fixation**: Fixate the distal leg against the treatment surface with a strap.
- **Therapist's moving hands:** With your left hand, grip around the talus and calcaneus from the fibular side; with your right hand, grip around the anterior talus and forefoot; position your right forearm in line with the treatment plane; maintain a Grade I distal traction.

Procedure

- Apply a Grade III posterior glide movement to the talus.

■ Dorsal flexion progression (not shown)

- Position the ankle near the end range-of-motion into dorsal flexion.

Talocrural posterior glide
for restricted dorsal flexion (cont'd)

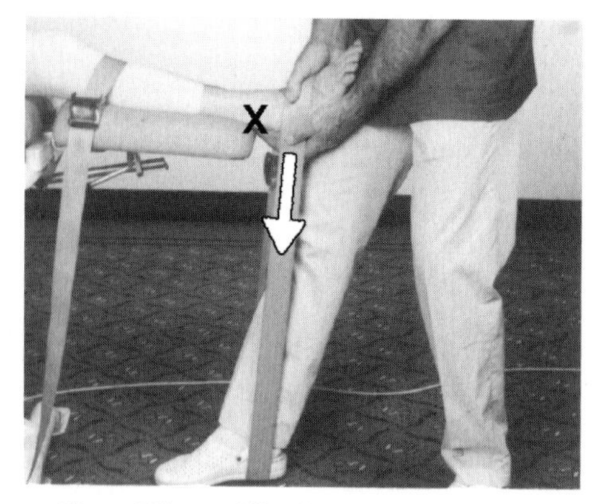

Figure 59b – mobilization in resting position

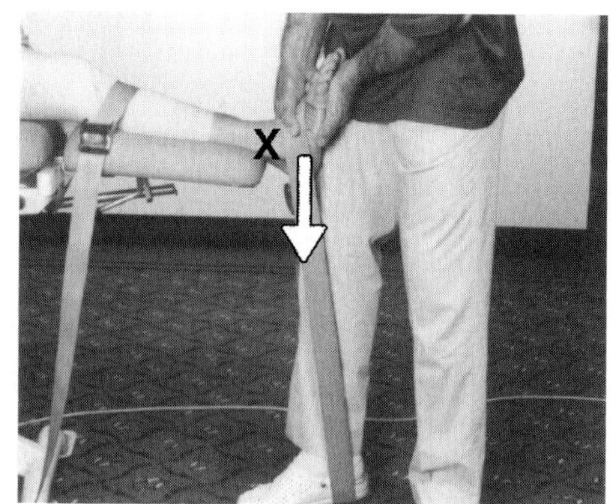

Figure 59c – mobilization in dorsal flexion

Foot

■ Figure 59b: Mobilization in resting position, alternate technique

Objective

- To increase foot dorsal flexion (Convex Rule).

Starting position

- The posterior side of the patient's leg rests on the treatment surface with the foot extended over the edge.
- Position the talocrural joint in its resting position.

Hand placement and fixation

- **Fixation**: The distal tibia and fibula are fixated against the treatment surface.
- **Therapist's moving hands:** With your right hand, grip the patient's midfoot from the tibial side with your web space over the dorsal talus; with your left hand, grip around the posterior talus; position your right forearm in line with the treatment plane; maintain a Grade I distal traction.

Procedure

- Apply a Grade III dorsal glide movement to the talus. For treatment of longer duration, use a traction strap: your heel applies the gliding force while your hands guide the movement.

■ Figure 59c: Dorsal flexion progression, alternate technique

- Position the ankle near the end range-of-motion into dorsal flexion.
- Apply a Grade III posterior glide movement to the talus.

CHAPTER 18

LEG

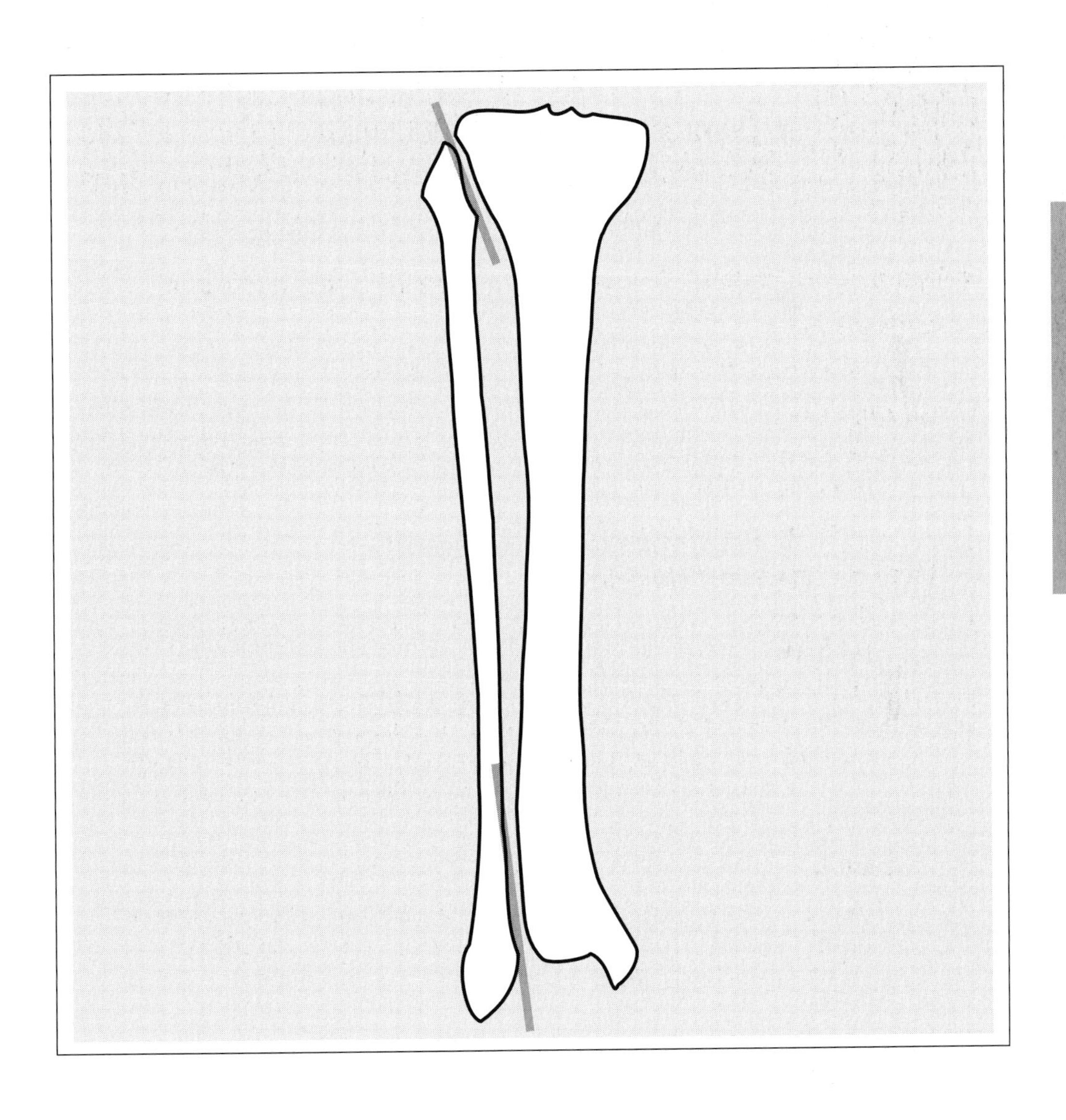

18 Leg

(crus)

Functional anatomy and movement

The leg consists of the tibia and fibula with the crural interosseous membrane.

Distal tibiofibular syndesmosis

(syndesmosis tibiofibularis distalis)

The distal tibiofibular syndesmosis has little movement.

"Long" syndesmosis

(syndesmosis tibiofibularis)

The "long" syndesmosis extends along the interosseous borders of the tibia and fibula with the interosseous membrane, and allows slight movement of the fibula in relation to the tibia.

Proximal tibiofibular joint

(art. tibiofibularis)

The proximal tibiofibular joint is an anatomically simple and mechanically compound joint and may communicate with the knee joint via the subpopliteal recess. According to MacConaill, the proximal tibiofibular joint is a sellar joint when it is considered as a functional unit with the distal syndesmosis.

Bony palpation

- Proximal end of the fibula (head of fibula)
- Proximal tibiofibular joint space
- Distal end of the fibula (lateral malleolus)
- Medial malleolus

Ligaments

Distal tibiofibular:

- Tibiofibular (anterior and posterior) ligaments
- Talofibular (anterior and posterior) ligaments

Proximal tibiofibular:

- Anterior and posterior ligaments

Bone movement and axes

Isolated active movements of the fibula do not take place. However, movement of the fibula does take place with foot and ankle movements.

- **Inversion of the foot:** The head of the fibula glides distally and slightly dorsally. This movement is considered to be lateral rotation by some authors.
- **Eversion of the foot:** The head of the fibula glides proximally and slightly ventrally (medial rotation).
- **Dorsal flexion of the ankle:** The fibula glides slightly proximally.
- **Plantar flexion of the ankle:** The fibula glides slightly distally.

End feel

- Firm

Joint movement (gliding)

- *Proximal tibiofibular joint:* Concave Rule — The articular surface of the head of the fibula is concave, while the articular surface of the tibial condyle is convex.

Treatment plane

- *Proximal tibiofibular joint:* on the concave joint surface of the head of the fibula

Zero position

- Same as for knee, foot and ankle

Resting position

- Approximately 10° ankle plantar flexion

Close-packed position

- Maximal dorsal flexion in the ankle joint

Leg examination scheme

(Refer to Chapters 3 and 4 for more information on examination)

Tests of function

1. **Active and passive movements, including stability tests and end-feel**
2. **Translatoric joint play movements, including end-feel**

   Gliding

   Distal tibiofibular syndesmosis (Figure 60a)

   Posterior-anterior glide of the lateral malleolus

   Proximal tibiofibular joint

   Anterior glide of the fibular head (Figure 61a)

   Posterior glide of the fibular head (Figure 61c)

3. **Resisted movements**

   (see *Ankle examination scheme* and *Knee examination scheme*)

4. **Passive soft tissue movements**

   Physiological

   Accessory

5. **Additional tests**

Trial treatment

Distal tibiofibular syndesmosis (Figure 60b)

Proximal tibiofibular joint (Figure 61b)

Leg

■ Leg techniques

Distal tibiofibular joint

Proximal tibiofibular joint

Distal fibula posterior glide

for hypomobility

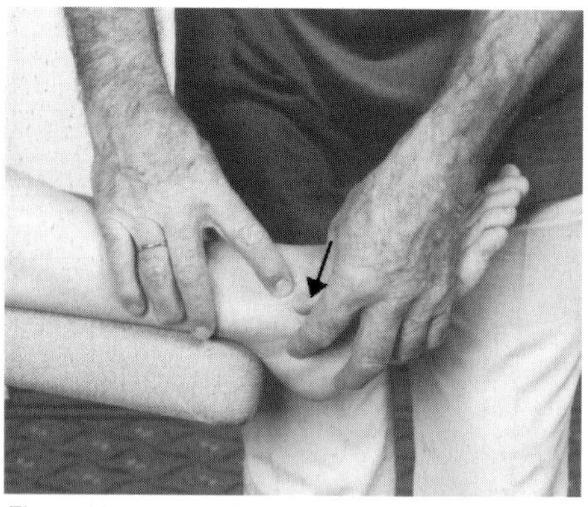
Figure 60a – test and mobilization in resting position

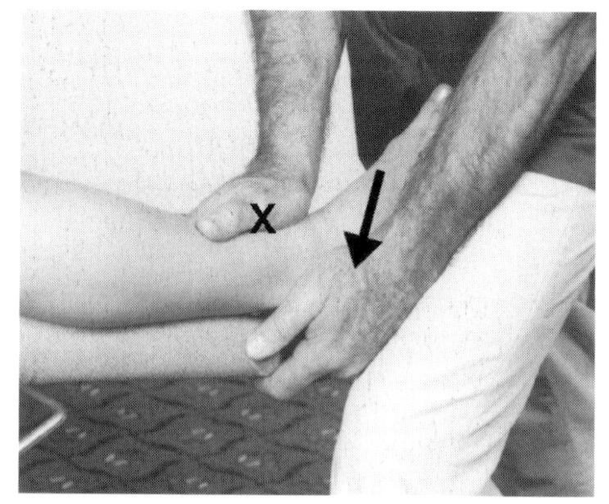

Figure 60b – mobilization in resting position

Leg

■ Figure 60a: Test and mobilization in resting position

Objective

- To evaluate the quantity and quality of posterior glide joint play in the tibiofibular syndesmosis, including end-feel.
- To increase tibiofibular mobility.

Starting position

- The posterior side of the patient's leg rests on the treatment surface, with the ankle off the edge in the resting position.

Hand placement and fixation

- **Fixation and therapist's stable hand (right):** The patient's tibia is fixated by the treatment surface; place your palpating finger in the joint space.
- **Therapist's moving hand (left):** Grip around the patient's lateral malleolus with your fingers and thumb.

Procedure

- Press the patient's lateral malleolus in a dorsal direction to apply a Grade II or III posterior glide movement to the fibula.

■ Figure 60b: Mobilization in resting position

- Apply a Grade III posterior glide movement to the fibula.
- Adapt the same procedure with the patient in prone for restricted fibular anterior glide.

Proximal fibula anterior glide
for hypomobility

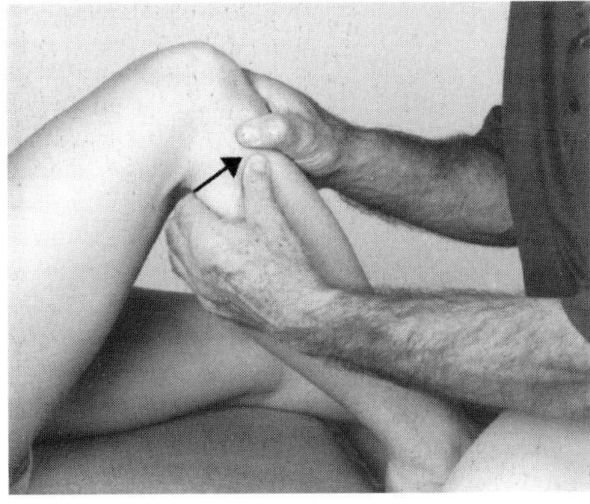
Figure 61a – test and mobilization in resting position

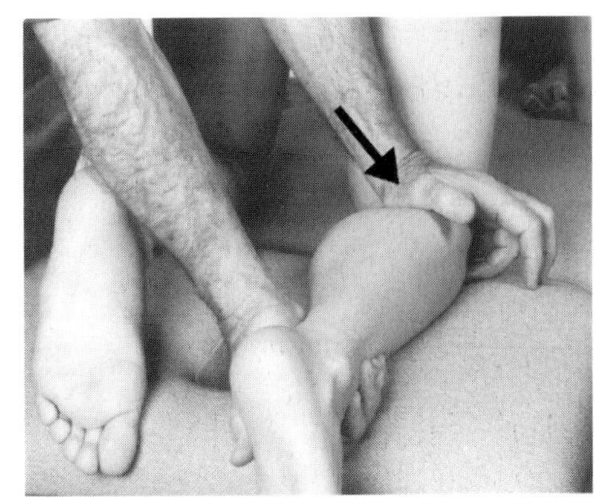
Figure 61b – mobilization in resting position

■ Figure 61a: Test and mobilization in resting position

Objective

- To evaluate the quantity and quality of anterior glide joint play in the proximal tibiofibular joint, including end-feel.
- To increase knee and ankle mobility.

Starting position

- The patient lies supine with the knee flexed.

Hand placement and fixation

- **Fixation and therapist's stable hand (right):** Hold around the patient's proximal tibia; enhance fixation of the tibia by sitting on the patient's foot.
- **Therapist's moving hand (left):** Grip around the patient's fibular head with your fingers and thumb.

Procedure

- Pull the fibular head in an anterior-lateral direction to apply a Grade II or III anterior glide movement in line with the treatment plane.

■ Figure 61b: Mobilization in resting position

- The patient is on their hands and knees with their foot over the end of the treatment surface; with your right hand, grip around the distal tibia; with your left hand and fingers; grip around the fibular head and apply a Grade III anterior-lateral glide movement.
- Be careful to avoid pressure on the fibular nerve.

Proximal fibula posterior glide

test and mobilization

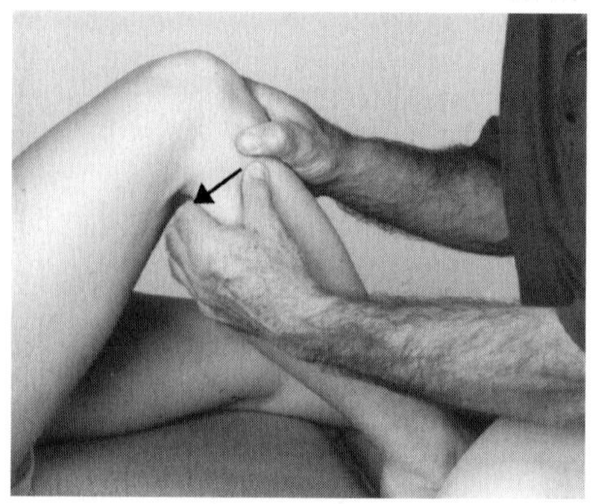

Figure 61c – test and mobilization in resting position

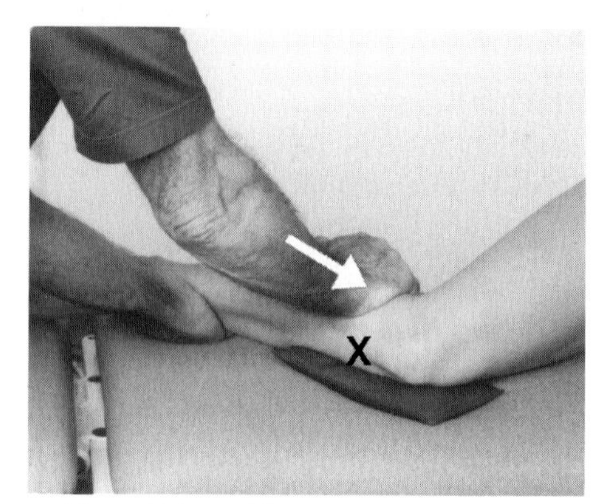

Figure 61d – mobilization in resting position

■ Figure 61c: Test and mobilization in resting position

Objective

- To evaluate the quantity and quality of posterior glide joint play in the proximal tibiofibular joint, including end-feel.
- To increase knee and ankle mobility.

Starting position

- The patient lies supine with the knee flexed.

Hand placement and fixation

- **Fixation and therapist's stable hand (right):** Hold around the patient's proximal tibia; enhance fixation of the tibia by sitting on the patient's foot.
- **Therapist's moving hand (left):** Grip around the patient's fibular head with your fingers and thumb.

Procedure

- Push the fibular head in an posterior-medial direction to apply a Grade II or III posterior glide movement in line with the treatment plane.

■ Figure 61d: Mobilization in resting position

- The tibial side of the patient's leg rests on the treatment surface; with your left hand, fixate the distal tibia against the treatment surface; with your right hand, grip around the fibular head.
- Apply a Grade III posterior-medial glide movement in line with the treatment plane; be careful to avoid pressure on the fibular nerve.

CHAPTER 19

KNEE

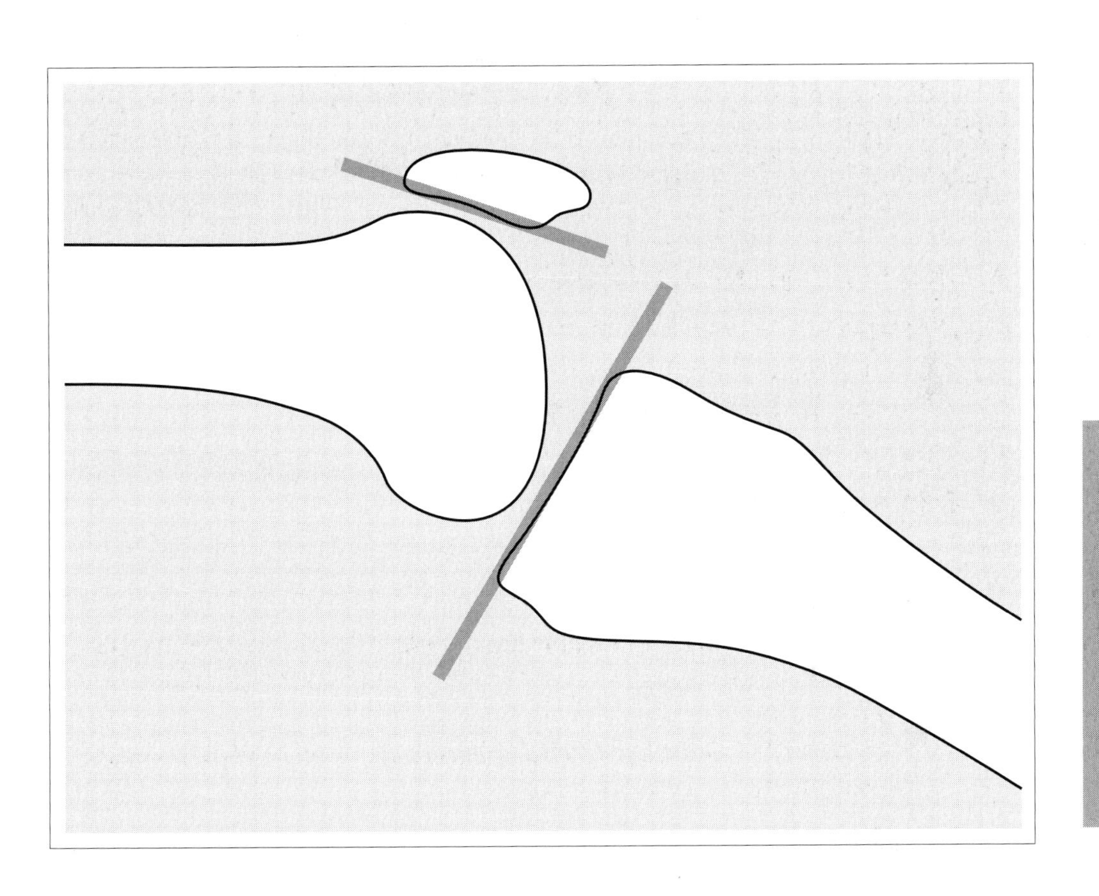

19 Knee

Functional anatomy and movement

Tibiofemoral joint

(art. genu)

The knee joint is an anatomically compound and mechanically simple biaxial joint (modified ovoid). It is mechanically compound when the tibiofibular joint is included. The distal end of the femur (convex surface) has two facets (medial and lateral condyles) for the menisci (semilunar cartilages). The proximal end of the tibia has two concave joint surfaces, separated by the intercondylar eminence for articulation with the menisci.

Patellofemoral joint

The distal end of the femur also has a convex facet for articulation with the concave patella.

Bony palpation

Anterior
- Femur
- Patella
- Knee joint space and menisci
- Tibial tuberosity
- Iliotibial tubercle

Medial
- Tibial plateau
- Medial femoral condyle
- Knee joint space
- Adductor tubercle

Lateral
- Lateral tibial plateau
- Lateral femoral condyle
- Knee joint space
- Fibular head

Posterior
- Medial femoral condyle
- Lateral femoral condyle
- Fibular head

Ligaments

- Anterior and posterior cruciate ligaments (intra-articular)
- Tibial collateral ligament (also, attaches to the medial meniscus and joint capsule)
- Fibular collateral ligament (attaches to the lateral part of the fibular head; not attached to the lateral meniscus or joint capsule)
- Patellar ligament (patellar tendon)
- Transverse ligament
- Meniscotibial (coronary) ligaments (medial and lateral). The medial meniscotibial ligament extends from the medial meniscus to the proximal medial aspect of the tibia and also to the tibial collateral ligament. The lateral meniscotibial ligament extends from the lateral meniscus to the proximal lateral aspect of the tibia; it has a greater laxity than the medial meniscotibial ligament.

Bone movement and axes

Tibiofemoral joint:

- Flexion - extension: around a tibiofibular axis through the femoral condyles
- Internal - external rotation between the menisci and tibia: around a longitudinal axis through the medial intercondylar eminence. Rotation is greatest at 90° flexion.
- Abduction - adduction (passive lateral movement): around an anterior-posterior axis through the femur. Abduction and adduction are greatest at 30° flexion.

  When the knee is in full extension, the collateral ligaments become taut. This provides stability to the knee in standing and limits rotation and lateral movements in the knee.

Patellofemoral joint:

- Proximal and distal gliding during knee flexion and extension

End feel

Tibiofemoral joint:

- Extension: firm
- Flexion: soft

Patellar-femoral joint

- Firm in all directions

Joint movement (gliding)

- Concave Rule

  During flexion and extension, a combined movement of rolling and gliding takes place between the femur and menisci. Simultaneously, the menisci are pushed slightly dorsally on the tibia during flexion and ventrally with extension.

  Isolated rotation can only occur in a normal knee when the knee is in some degree of flexion. During knee extension, the tibia rotates into external rotation.

  The patella glides proximally on the femur during knee extension due to the quadriceps femoris contraction.

Treatment plane

- *Tibiofemoral joint:* on the concave joint surface of the tibia
- *Patellofemoral joint:* on the concave posterior joint surface of the patella

Zero position

- The longitudinal axes passing through the femur and tibia meet in the frontal plane and form a laterally facing angle of approximately 170° (valgus).

Resting position

- 25° to 40° flexion

Close-packed position

- Maximal knee extension

Capsular pattern

- Flexion - extension: The proportion of these limitations is such that, with 90° of limited flexion, there is only 5° of limited extension.
- Rotation is limited only when there is marked limitation of both flexion and extension.

Knee examination scheme

(Refer to Chapters 3 and 4 for more information on examination)

Tests of function

1. Active and passive movements, including stability tests and end-feel

Flexion	160°
Extension from zero	5°
Lateral rotation at 90° knee flexion	45°
Medial rotation at 90° knee flexion	15°
Abduction (passive)	
Adduction (passive)	
Stability tests (passive)	(Figure 62a, b, c, d)
Meniscus tests (passive)	(Figure 63a, b, c, d)

2. Translatoric joint play movement, including end-feel

Traction - compression	(Figure 64a)
Gliding	
Posterior - anterior	(Figure 66a, b)
Lateral	(Figure 69)
Medial	(Figure 70)
Patella distal	(Figure 71a)

3. Resisted movements

	OTHER FUNCTIONS
Flexion	
Biceps femoris	Lateral rotation
Semitendinosus	Medial rotation
Semimembranosus	Medial rotation
Gastrocnemius	Ankle plantar flexion
Popliteus	Medial rotation
Extension	
Rectus femoris	Hip flexion
Vastus lateralis, medialis, intermedius	
Lateral Rotation	
Tensor fasciae latae	Extension
Biceps femoris	Flexion
Medial Rotation	
Sartorius	Flexion; hip lateral rotation
Gracilis	Flexion; hip medial rotation
Semitendinosus	Hip adduction, extension
Semimembranosus	Hip adduction, extension
Popliteus	Flexion

4. Passive soft tissue movements

Physiological
Accessory

5. Additional tests

Trial treatment

Traction	(Figure 64b)

Knee

Knee techniques

Knee

Patella

Knee lateral stability
test

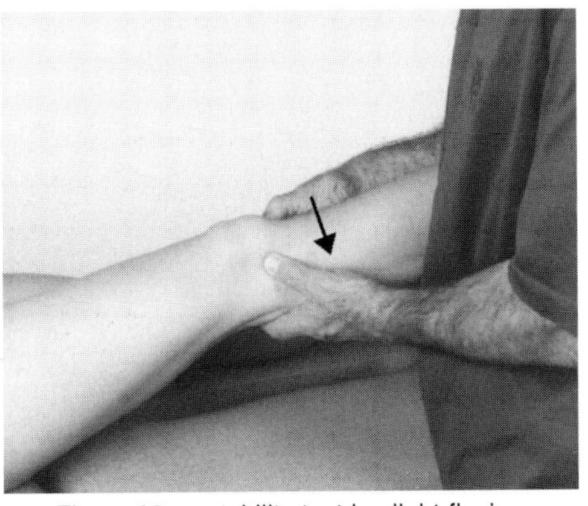
Figure 62a – stability test in slight flexion

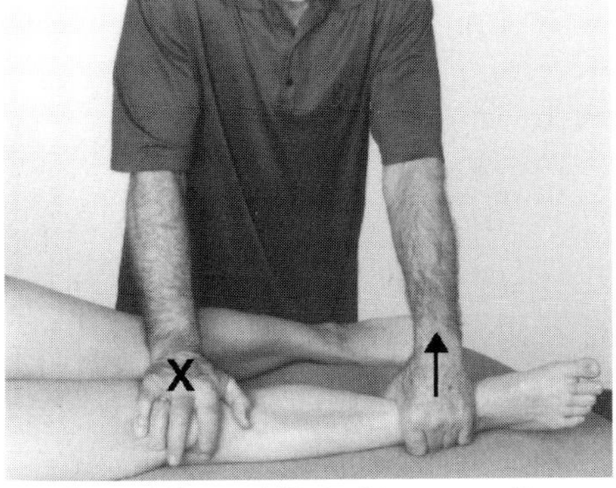

Figure 62b – stability test in zero position

■ Figure 62a: Stability test in slight flexion

Objective

- To test for ligamentous and capsular integrity in the lateral compartment of the knee.
- To evaluate the quantity and quality of lateral joint play in the knee, including hypermobility.

Starting position

- The patient is supine with the knee slightly flexed and the lower leg beyond the edge of the treatment surface.

Hand placement and fixation

- **Fixation:** Grip the patient's lower leg between your body and upper arm.
- **Therapist's moving hands:** Grip the patient's knee from both sides.

Procedure

- Apply a Grade II or III lateral movement to the knee; test for end-feel gently and with caution; the patient's thigh should move slightly during the test.

■ Figure 62b: Stability test in the zero position

- The lateral side of the patient's leg rests on the treatment surface; with your right hand, fixate the patient's knee against the treatment surface; with your left hand, grip the patient's distal leg and move it in a medial direction to produce lateral gapping. Tests ligamentous and capsular integrity in the lateral compartment of the knee
- The stability test is normal if the end-feel is very firm (i.e., firm "+"), there is very little movement, and there is no pain.

Knee medial stability
test

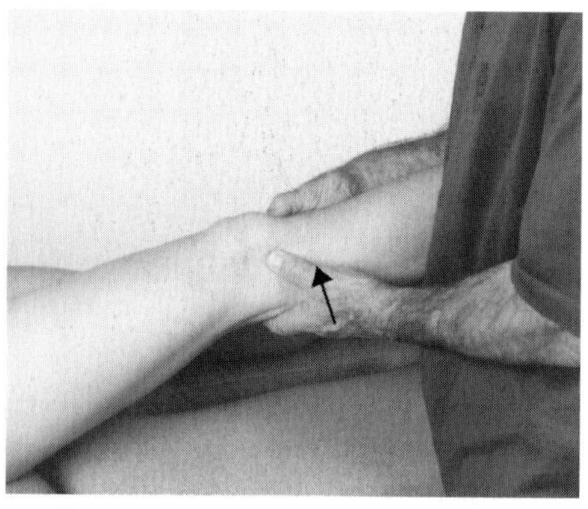

Figure 62c – stability test in slight flexion

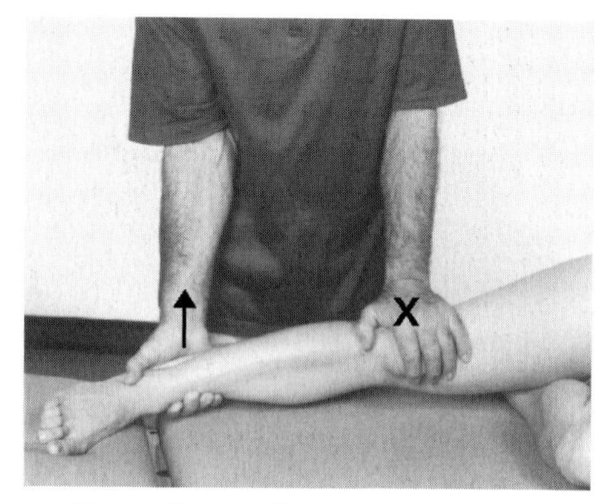

Figure 62d – stability test in zero position

■ Figure 62c: Stability test in slight flexion

Objective

- To test for ligamentous and capsular integrity in the medial compartment of the knee.
- To evaluate the quantity and quality of medial joint play in the knee, including hypermobility.

Starting position

- The patient is supine with the knee slightly flexed and the lower leg beyond the edge of the treatment surface.

Hand placement and fixation

- **Fixation:** Grip the patient's lower leg between your body and upper arm.
- **Therapist's moving hands:** Grip the patient's knee from both sides.

Procedure

- Apply a Grade II or III medial movement to the knee; test for end-feel gently and with caution; the patient's thigh should move slightly during the test.

■ Figure 62d: Stability test in the zero position

- The medial side of the patient's leg rests on the treatment surface; with your left hand, fixate the patient's knee against the treatment surface; with your right hand, grip the patient's distal leg and move it in a lateral direction to produce medial gapping. Tests ligamentous and capsular integrity in the medial compartment of the knee.
- The stability test is normal if the end-feel is very firm (i.e., firm "+"), there is very little movement, and there is no pain.

Knee meniscus
test in adduction

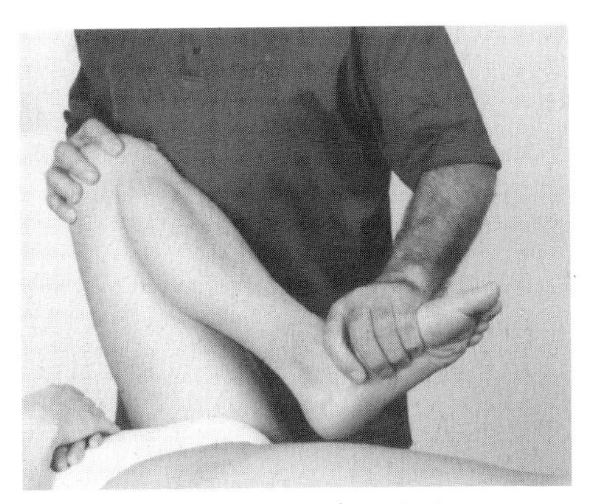

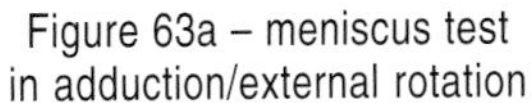

Figure 63a – meniscus test in adduction/external rotation

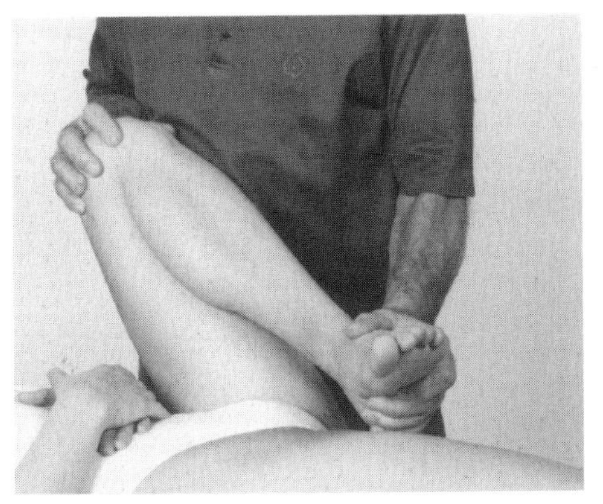

Figure 63b – meniscus test in adduction/internal rotation

■ Figure 63a: Meniscus test in adduction/external rotation

Objective

- To test for integrity of the knee meniscus. The test is positive if there are joint noises (e.g., snapping, popping), restricted movement, or pain.

Starting position

- The patient is supine with the knee and hip in the resting position (not shown).

Hand placement and fixation

- **Therapist's hand placement:** With your right hand, grip the patient's knee from the anterior side to guide and control hip position; palpate in the medial joint space; with your left hand, grip the patient's foot.

Procedure

- Adduct and externally rotate the lower leg; gently maintain the adduction/external rotation force while you move the knee into full flexion (pictured) and back again into full extension; the hip should not adduct nor abduct during the test.

Note

- With positive findings, refer the patient to an orthopedist for further examination.

■ Figure 63b: Meniscus test in adduction/internal rotation

- Adduct and internally rotate the lower leg; gently maintain the adduction/ internal rotation force while you move the knee into full flexion and back again into full extension.

Knee meniscus
test in abduction

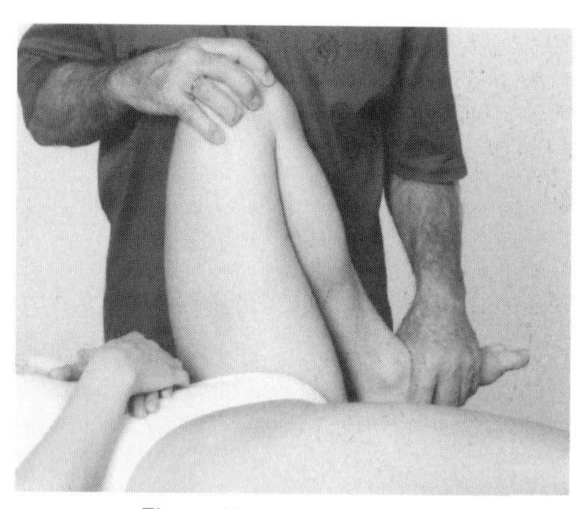

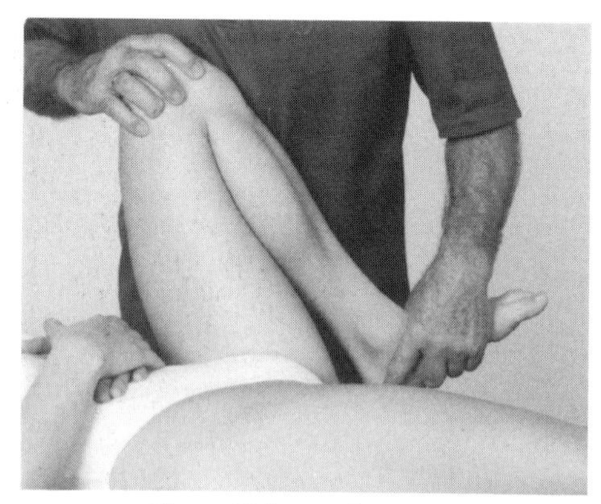

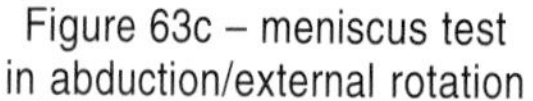
Figure 63c – meniscus test in abduction/external rotation

Figure 63d – meniscus test in abduction/internal rotation

■ Figure 63c: Meniscus test in abduction/external rotation

Objective

- To test for integrity of the knee meniscus. The test is positive if there are joint noises (e.g., snapping, popping), restricted movement or pain.

Starting position

- The patient is supine with the knee and hip in the resting position (not shown).

Hand placement and fixation

- **Therapist's hand placement:** With your right hand, grip the patient's knee from the anterior side to guide and control hip position; palpate in the medial joint space, with your left hand, grip the patient's foot.

Procedure

- Abduct and externally rotate the lower leg; gently maintain the abduction/external rotation force while you move the knee into full flexion (pictured) and back again into full extension; the hip should not adduct nor abduct during the test.

Note

- With positive findings, refer the patient to an orthopedist for further examination.

■ Figure 63d: Meniscus test in abduction/internal rotation

- Abduct and internally rotate the lower leg; gently maintain the abduction/internal rotation force while you move the knee into full flexion and back again into full extension.

Knee traction
for pain and hypomobility

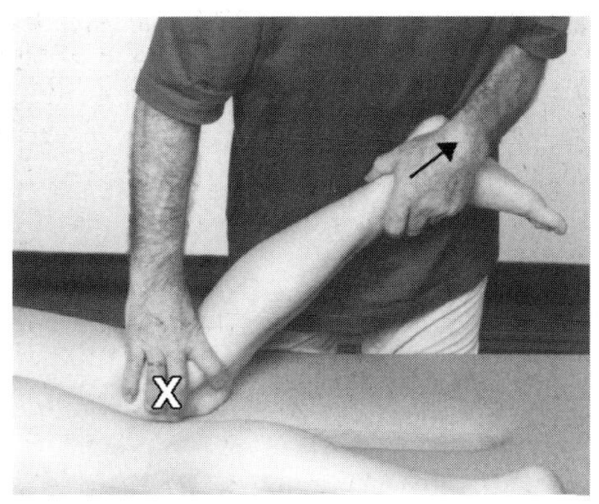

Figure 64a – test and mobilization in resting position

■ Figure 64a: Test and mobilization in resting position

Objective

- To evaluate the quantity and quality of traction joint play in the knee, including end-feel.
- To decrease pain or increase knee mobility.

Starting position

- The patient lies prone with the anterior side of the thigh on the treatment surface.
- Position the knee in its resting position.

Hand placement and fixation

- **Therapist's stable hand (right):** Fixate the patient's distal thigh against the treatment surface; place your palpating finger in the joint space.
- **Therapist's moving hand (left):** Grip the patient's leg above the ankle joint; position your forearm in line with the patient's lower leg.

Procedure

- Apply a Grade I, II, or III traction movement in line with the lower leg.

Knee traction
for hypomobility

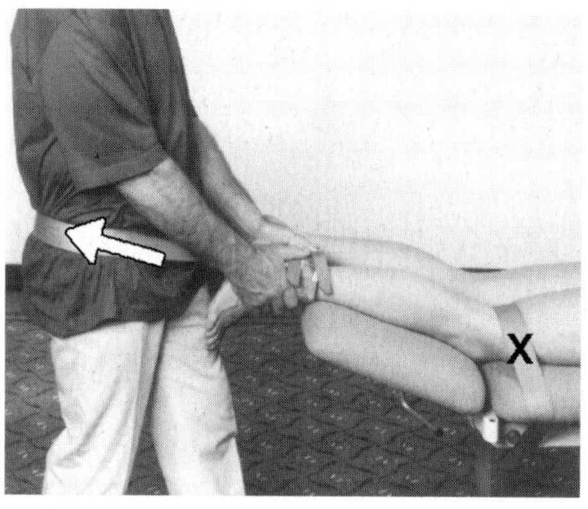

Figure 64b – mobilization in resting position

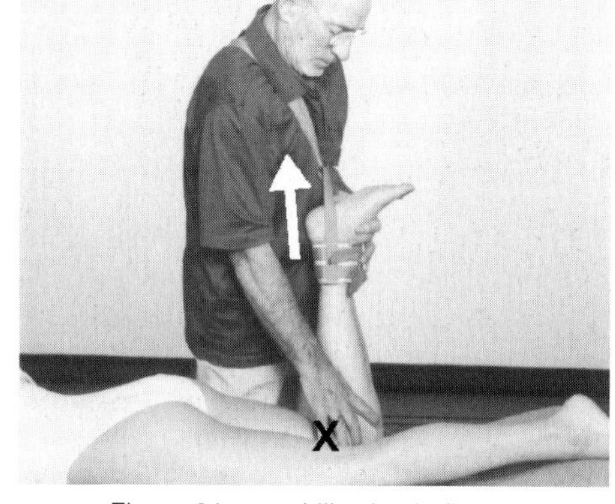

Figure 64c – mobilization in flexion

■ Figure 64b: Mobilization in resting position

Objective

- To increase knee mobility.

Starting position

- The patient lies prone with the anterior side of the thigh on the treatment surface.
- Position the knee in its resting position.

Hand placement and fixation

- **Fixation:** Fixate the patient's distal thigh against the treatment surface with a strap.
- **Therapist's moving hands:** Grip above the patient's ankle joint with both hands; for longer treatments, enhance your grip by using a traction cuff and/or a strap around your body and hands.

Procedure

- Apply a Grade III traction movement in line with the lower leg by shifting your body backward.

■ Figure 64c: Flexion progression

- Position the joint close to its end range-of-motion in flexion; adapt the technique by positioning the traction strap over your shoulder
- Apply a Grade III traction movement by bending and extending your knees; palpate in the joint space.

Knee traction

for hypomobility (alternate technique)

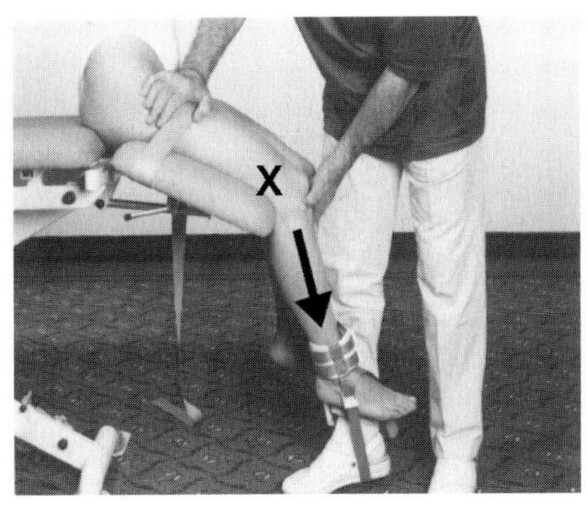

Figure 65a – mobilization in resting position

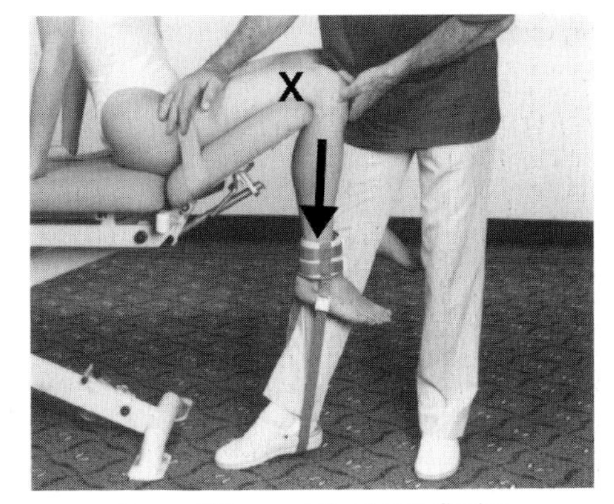

Figure 65b – mobilization in flexion

■ Figure 65a: Mobilization in resting position

Objective

- To increase knee mobility.

Starting position

- The patient sits on the treatment table with the knee over the edge.
- Position the knee in its resting position.

Hand placement and fixation

- **Fixation and therapist's stable hand (right):** Fixate the patient's proximal thigh against the treatment surface with your hand and a strap; the patient's distal thigh is fixated by the edge of the treatment surface.
- **Therapist's moving hand (left):** Grip below the patient's knee; attach a traction strap above the patient's ankle with its stirrup adjusted just above the floor; put your foot in the stirrup.

Procedure

- Apply a Grade III traction movement in line with the lower leg by pressing the traction stirrup down with your heel; keep your forefoot on the floor; palpate in the joint space.

■ Figure 65b: Flexion progression

- Position the joint close to its end range-of-motion in flexion. Control the position of the knee by raising or lowering the end of the treatment table. Apply a Grade III traction movement to the lower leg.

Knee posterior and anterior glide
test

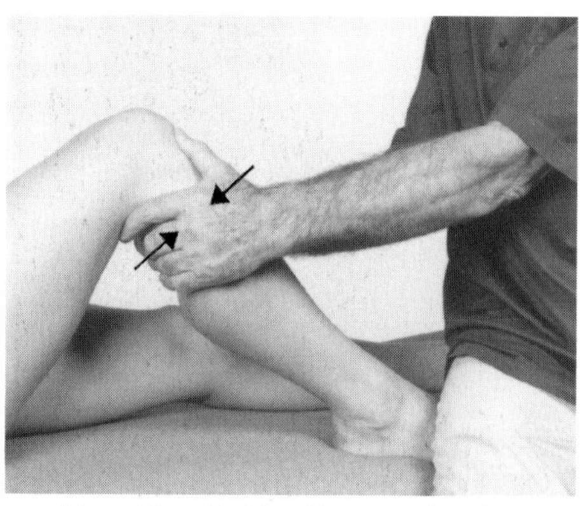

Figure 66a – test in mid-range-of-motion

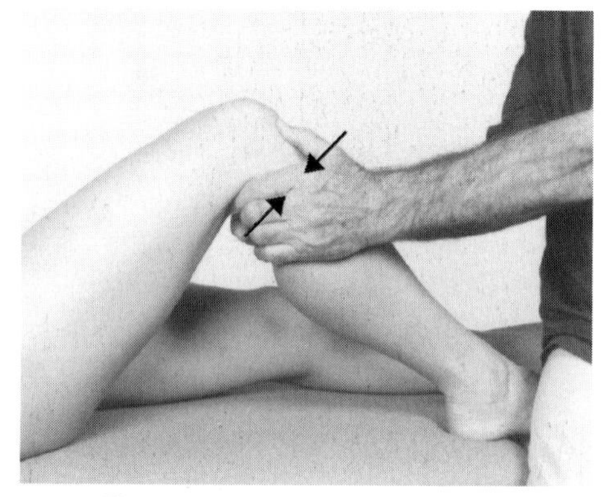

Figure 66b – test in internal rotation

■ Figure 66a: Test in mid-range-of-motion

Objective

- To evaluate the quantity and quality of posterior and anterior glide joint play in the knee, including end-feel. Restricted posterior glide is associated with restricted knee flexion; restricted anterior glide is associated with restricted knee extension.

Starting position

- The patient lies supine.
- Position the knee at about 90° flexion. This technique is difficult to perform with the patient's knee in the resting position.

Hand placement and fixation

- **Fixation:** No external fixation of the femur is necessary; the foot is fixated on the treatment surface by the weight of the patient's thigh; enhance fixation of the patient's foot by sitting on it.
- **Therapist's moving hands:** Grip below the patient's knee with both hands; palpate the joint space with your thumbs.

Procedure

- Apply a Grade II or III posterior or anterior glide movement to the tibia by leaning through your extended arms and shifting your body forward and backward.

■ Figure 66b: Test in internal rotation

- Position the joint close to its end range-of-motion in internal rotation.
- **Ligamentous stability test**: Posterior glide tests the posterior cruciate; anterior glide tests the anterior cruciate. With intact cruciate ligaments, range-of-motion into posterior and anterior glide is greater with the lower leg internally rotated than when the leg is in a neutral position or externally rotated.

Knee posterior glide
for restricted flexion

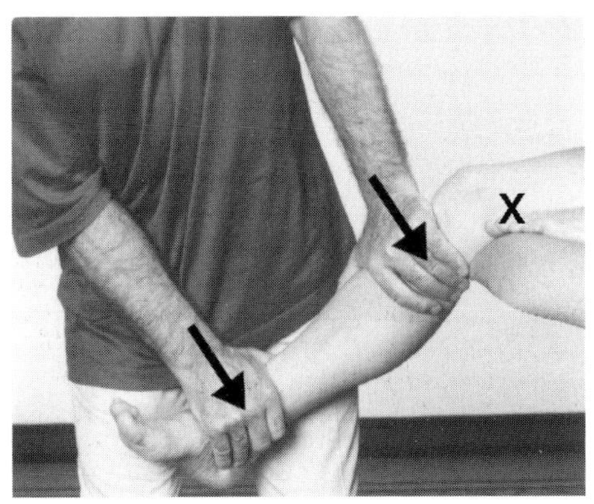

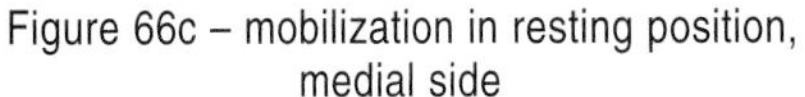
Figure 66c – mobilization in resting position, medial side

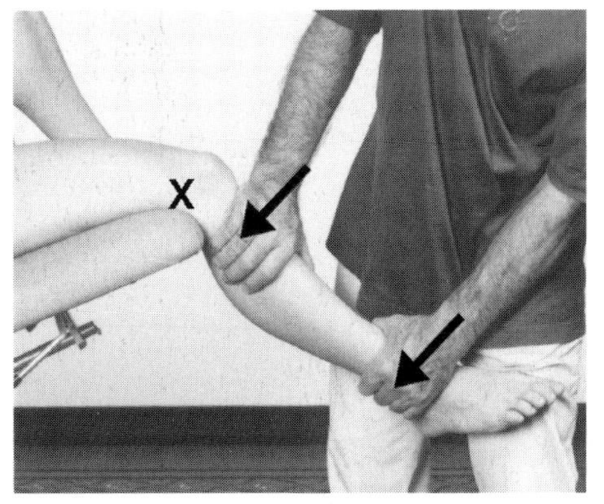

Figure 66d – mobilization in resting position, lateral side

■ Figure 66c: Mobilization in resting position, medial side

Objective

- To increase knee flexion (Concave Rule) and internal rotation range-of-motion.

Starting position

- The patient sits or lies with the lower leg over the edge of the treatment surface.
- Position the knee in its resting position.

Hand placement and fixation

- **Fixation:** The patient's thigh is fixated against the treatment surface.
- **Therapist's moving hands:** Hold the lower leg from the anterior-medial side with both hands; grip your right hand above the ankle and your left hand below the knee.

Procedure

- Apply a Grade III posterior glide movement to the medial tibia by leaning through your extended arms and bending your knees.

Note

- Knee mobilization is easier and more effective when treatment is directed specifically to the medial or lateral side.

■ Figure 66d: Mobilization in resting position, lateral side

- Hold the lower leg from the anterior-lateral side.
- Apply a Grade III posterior glide movement to increase knee flexion (Concave Rule) and external rotation range-of-motion.

Knee posterior glide
for restricted flexion (supine)

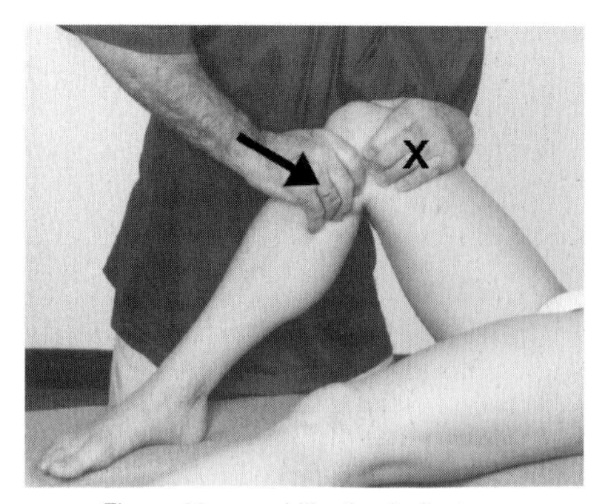

Figure 66e – mobilization in flexion, medial side

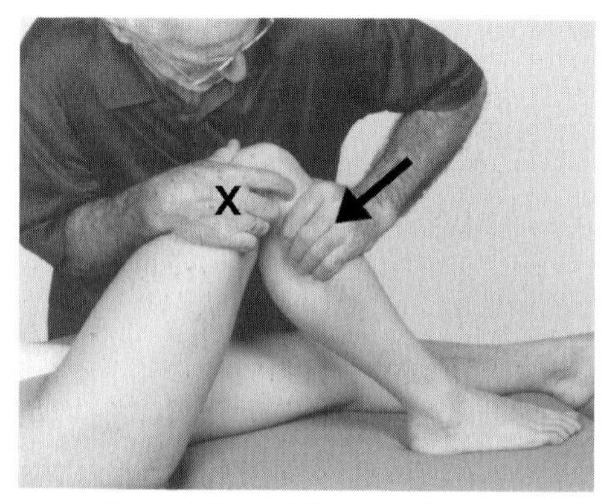

Figure 66f – mobilization in flexion, lateral side

■ Figure 66e: Flexion progression, medial side

Objective

- To increase knee flexion (Concave Rule) and internal rotation range-of-motion.

Starting position

- The patient is supine.
- Position the knee near its end range-of-motion into flexion-internal rotation.

Hand placement and fixation

- **Therapist's stable hand (left):** Hold the patient's thigh above the knee and fixate it against your body.
- **Therapist's moving hand (right):** Grip the lower leg below the knee from the anterior-medial side.

Procedure

- Apply a Grade III posterior glide movement to the medial tibia.

Note

- Knee mobilization is easier and more effective when treatment is directed specifically to the medial or lateral side.

■ Figure 66f: Flexion progression, lateral side

- Position the knee near its end range-of-motion into flexion-external rotation.
- Hold the lower leg from the anterior-lateral side.
- Apply a Grade III posterior glide movement to the tibia.
- To increase knee flexion (Concave Rule) and external rotation range-of-motion.

Knee posterior glide
for restricted flexion (prone)

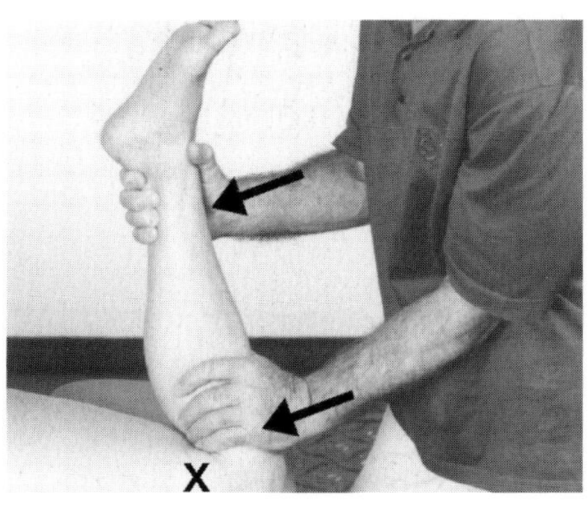

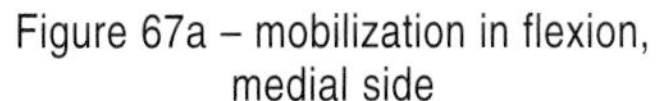
Figure 67a – mobilization in flexion, medial side

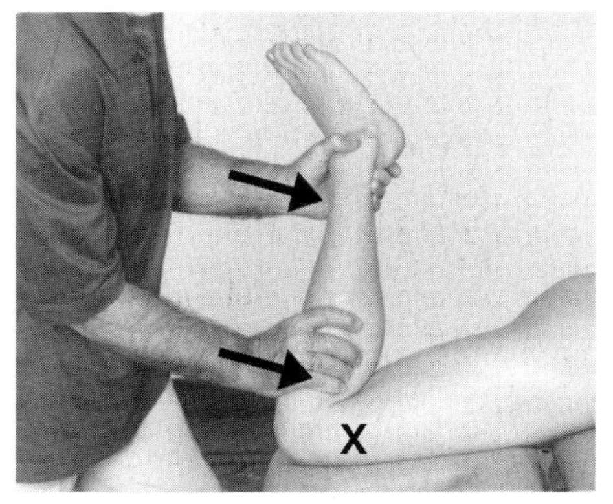

Figure 67b – mobilization in flexion, lateral side

■ Figure 67a: Flexion progression, medial side (alternate technique)

Objective

- To increase knee flexion (Concave Rule) and internal rotation range-of-motion.

Starting position

- The patient lies prone with the knee near the edge of the treatment table.
- Position the knee near its end range-of-motion into flexion-internal rotation

Hand placement and fixation

- **Fixation:** The patient's thigh is fixated against the treatment surface.
- **Therapist's moving hands:** Hold the lower leg from the anterior side with both hands; grip your right hand above the ankle and your left hand below the knee with your hypothenar eminence on the medial tibia; brace your left arm against your body.

Procedure

- Apply a Grade III posterior glide movement to the medial tibia by leaning through your left forearm; move both your hands and body together as one.

Note

- Knee mobilization is easier and more effective when treatment is directed specifically to the medial or lateral side.

■ Figure 67b: Mobilization in flexion, lateral side (alternate technique)

- Position the knee near its end range-of-motion into flexion-external rotation.
- Apply a Grade III posterior glide movement to the tibia.
- To increase knee flexion (Concave Rule) and external rotation range-of-motion.

Knee anterior glide
for restricted extension

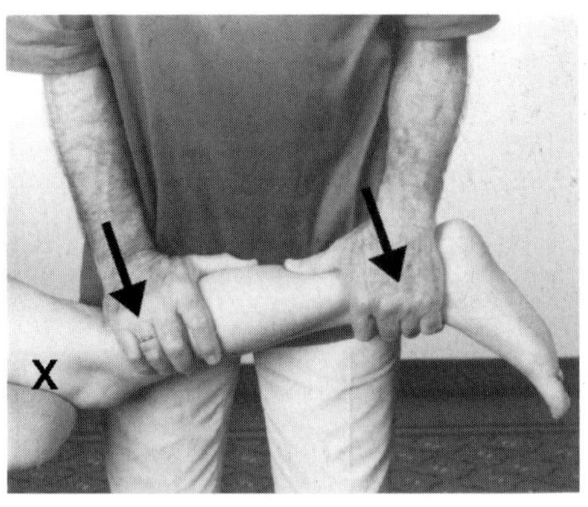

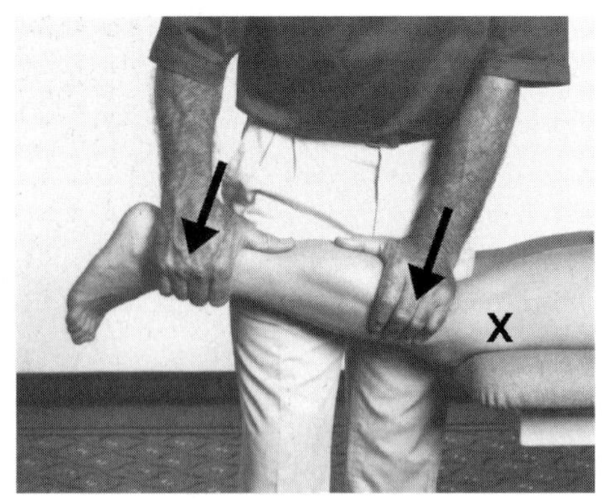

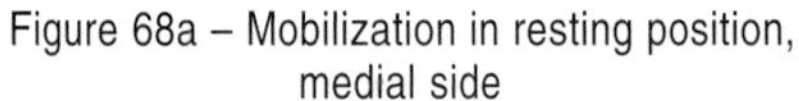

Figure 68a – Mobilization in resting position, medial side

Figure 68b – mobilization in resting position, lateral side

■ Figure 68a: Mobilization in resting position, medial side

Objective

- To evaluate the quantity and quality of anterior glide joint play of the proximal medial tibia.

Starting position

- The patient lies prone with the knee near the edge of the treatment table.
- Position the knee in its resting position.

Hand placement and fixation

- **Fixation:** The patient's thigh is fixated against the treatment surface.
- **Therapist's moving hands:** Hold the patient's lower leg against your body with both hands; grip from the medial side with your left hand proximal to the ankle and your right hand distal to the knee; place your right hypothenar eminence on the tibia.

Procedure

- Apply a Grade III anterior glide movement to the proximal medial tibia by leaning through your right forearm and bending your knees; move both your hands and body together as one.

Note

- Knee mobilization is easier and more effective when treatment is directed specifically to the medial or lateral side.

■ Figure 68b: Mobilization in resting position, lateral side

- Grip from the lateral side with your right hand proximal to the ankle and your left hand distal to the knee; place your left hypothenar eminence on the lateral tibia.
- Apply a Grade III anterior glide movement to the proximal lateral tibia to increase knee extension (Concave Rule) and internal rotation range-of-motion by increasing lateral tibial anterior glide.

Knee lateral glide

for restricted flexion and extension

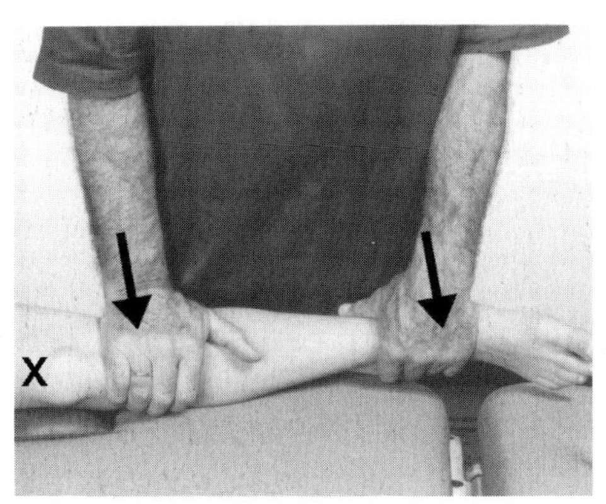

Figure 69 – test and mobilization in resting position

■ Figure 69: Test and mobilization in resting position

Objective

- To evaluate the quantity and quality of lateral glide joint play in the knee, including end-feel.
- To increase knee flexion, extension and rotation range-of-motion.

Starting position

- The patient is side-lying with the lateral side of the leg on the treatment surface.
- Position the knee in its resting position.

Hand placement and fixation

- **Fixation:** The patient's distal thigh is fixated by the treatment surface; to enhance the fixation, place a sandbag or wedge just proximal to the joint space.
- **Therapist's moving hands:** Hold the patient's lower leg from the medial side with both hands; with your left hand, grip proximal to the ankle; with your right hand, grip distal to the knee with your hypothenar eminence on the medial tibia just distal to the joint space.

Procedure

- Apply a Grade II or III lateral glide movement by leaning your body through your extended arms; move both hands and your body together as one.

■ Flexion and extension progression (not shown)

- Position the knee near its end range-of-motion into flexion (for restricted flexion) or extension (for restricted extension).
- Apply a Grade II or III lateral glide movement to the medial proximal tibia.

Knee medial glide
for restricted flexion and extension

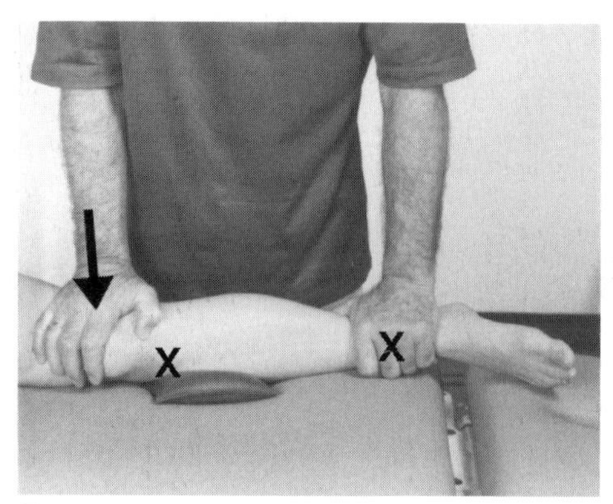

Figure 70 – test and mobilization in resting position

■ Figure 70: Test and mobilization in resting position

Objective

- To evaluate the quantity and quality of medial glide joint play in the tibia, including end-feel.
- To increase knee flexion and extension range-of-motion.

Starting position

- The patient is side-lying with the lateral side of the leg on the treatment surface.
- Position the knee in its resting position.

Hand placement and fixation

- **Fixation and therapist's stable hand (left):** Place a sandbag or wedge just distal to the knee; grip proximal to the patient's ankle and fixate it against the treatment surface.
- **Therapist's moving hand (right):** Grip around the medial side of the patient's thigh just proximal to the knee.

Procedure

- Apply a relative medial glide movement to the tibia by performing a Grade II or III lateral glide movement to the femur; lean your body through your extended right arm to produce the movement.

■ Flexion and extension progression (not shown)

- Position the knee near its end range-of-motion into flexion (for restricted flexion) or extension (for restricted extension).
- Apply a relative medial glide movement to the tibia by applying a Grade III lateral glide movement to the femur.

Patella distal glide
for restricted flexion

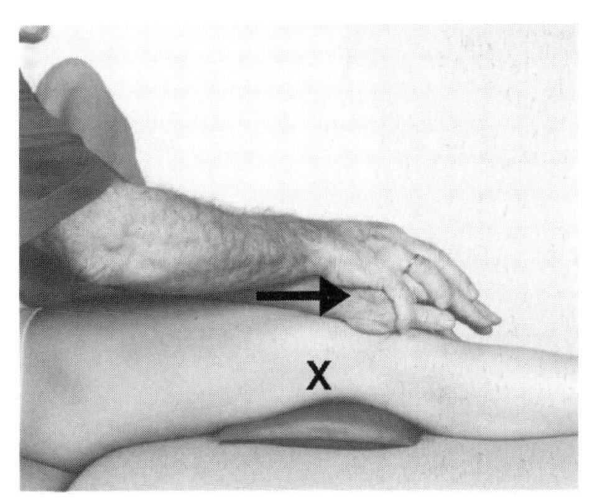

Figure 71a – test and mobilization in resting position

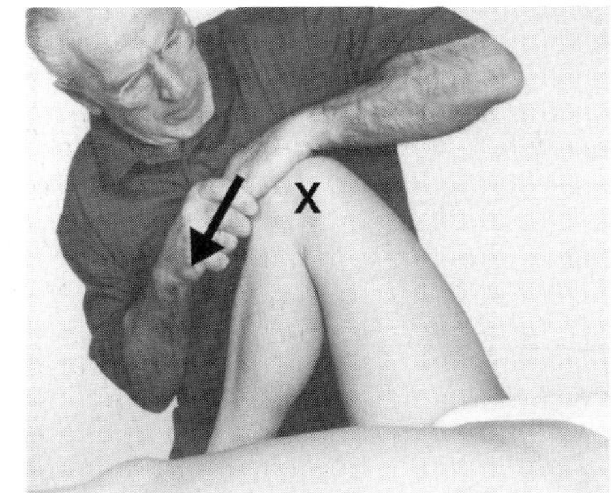

Figure 71b – mobilization in flexion

■ Figure 71a: Test and mobilization in resting position

Objective

- To evaluate the quantity and quality of patella distal glide joint play.
- To increase knee flexion range-of-motion by increasing patellar distal glide.

Starting position

- The patient lies supine.
- Use a sandbag to position the knee in its actual resting position; adjust the size of the sandbag to control the amount of knee flexion.

Hand placement and fixation

- **Fixation:** The knee is fixated by the sandbag.
- **Therapist's moving hands:** Grip with the heel of your left hand over the proximal edge of the patella and your fingers around the distal aspect of the patella; rest your left forearm along the patient's thigh; place your right hand over your left hand to enhance your grip.

Procedure

- Apply a Grade II or III distal glide movement to the patella; keep your forearms parallel to the thigh to avoid dorsally-directed compression forces to the patella; use your grip on the patella to simultaneously apply a Grade I traction to avoid pain during the movement.

Note

- Restricted distal glide of the patella can be associated with restricted knee flexion. In this case, it is important to increase patellar mobility before using knee joint mobilization techniques.

■ Figure 71b: Flexion progression

- Position the joint close to its end range-of-motion in flexion; apply a Grade III distal glide movement to the patella.

Patella medial and lateral glide
for hypomobility

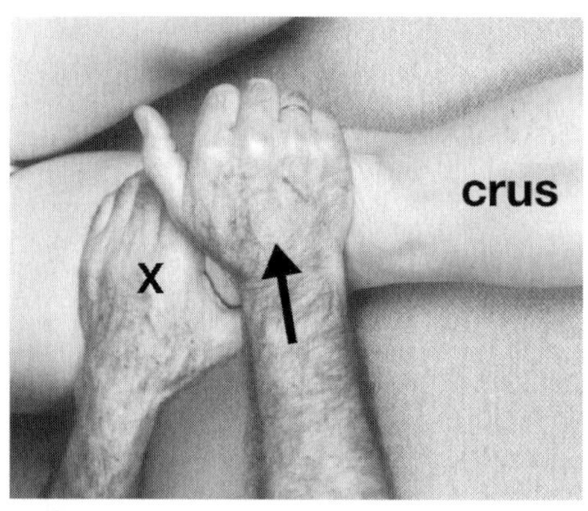

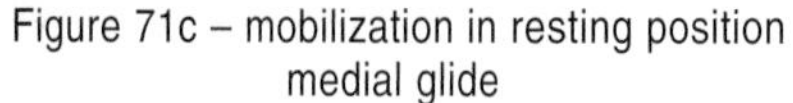
Figure 71c – mobilization in resting position medial glide

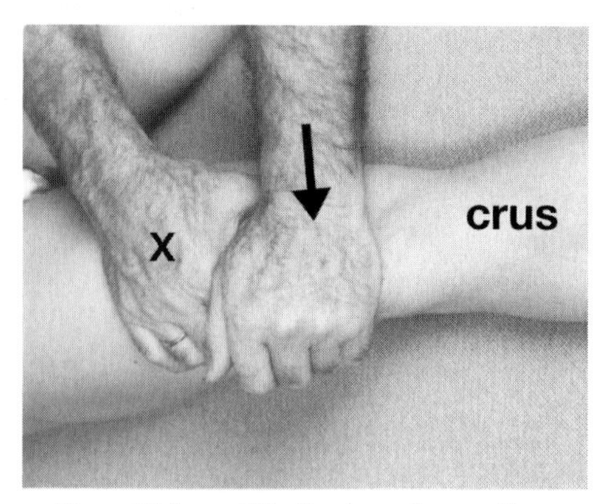

Figure 71d – mobilization in resting position lateral glide

■ Figure 71c: Mobilization in resting position, medial glide

Objective

- To increase knee movement by increasing medial glide of the patella.

Starting position

- The patient lies supine.
- Position the knee in its resting position. If necessary, use a sandbag to position the knee in its actual resting position.

Hand placement and fixation

- **Therapist's stable hand (left):** Grip around the thigh just above the patella from the anterior side.
- **Therapist's moving hand (right):** Grip with the heel of your hand over the lateral edge of the patella, with your forearm parallel to the treatment surface.

Procedure

- Apply a Grade II or III medial glide movement to the patella; keep your forearm parallel to the treatment surface to avoid dorsally directed compression forces to the patella.

■ Figure 71d: Mobilization in resting position, lateral glide

- Adapt the same technique.
- Apply Grade II or III lateral glide movement to the patella.

CHAPTER 20

HIP

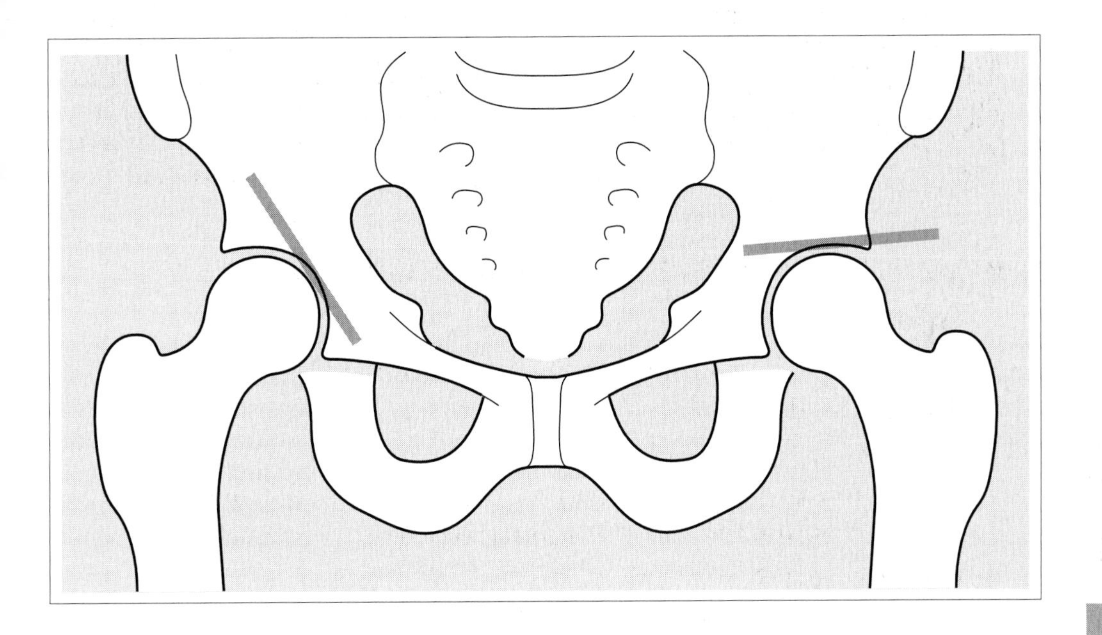

20 Hip

(art. coxae)

Functional anatomy and movement

The hip joint is an anatomically and mechanically simple, triaxial ball-and-socket joint (unmodified sellar).

The convex articulating surface is formed by the head of the femur (caput ossis femoris). The head of the femur is about two-thirds of a sphere on the neck of the femur (collum ossis femoris), which is itself approximately five centimeters long. The neck of the femur forms an angle of approximately 126° with the longitudinal axis of the femoral shaft (angle of declination) and an angle of approximately 12° with the frontal plane (angle of femoral torsion). The head of the femur faces the acetabulum in a medial, cranial, and slightly ventral direction.

The ilium contains the acetabulum (a lunate-shaped, concave articular surface) and the non-articular floor of the cavity, the acetabular fossa. The acetabular labrum is continuous with the acetabular rim; the transverse acetabular ligament running across the acetabular notch completes the circle. This ligament converts the notch into a foramen through which vessels and nerves pass in the ligament of the head of the femur.

The acetabular fossa is occupied by an articular fat pad (pulvinar acetabuli, or corpus adiposum fossae acetabuli), which can be pushed out or sucked in through the acetabular notch by variations in pressure.

Bony palpation

Ventral
- Hip joint
- Anterior superior iliac spine
- Anterior inferior iliac spine
- Iliac crest
- Symphysis pubis
- Lesser trochanter

Dorsal
- Iliac crest
- Posterior superior iliac spine
- Posterior inferior iliac spine
- Ischial tuberosity

Lateral
- Iliac crest
- Greater trochanter

Figure 20.1
Ligamentum capitis femoris

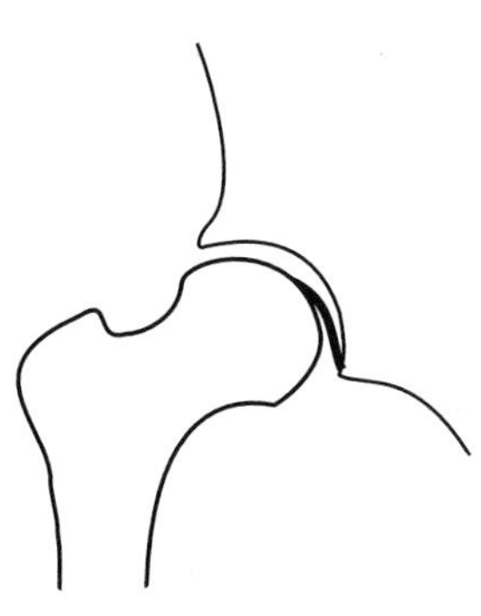

Ligaments

- Iliofemoral ligament (ventral)
- Pubofemoral ligament (caudal)
- Ischiofemoral ligament (dorsal/cranial)
- Zona orbicularis
- Ligament of the head of the femur (lig. capitis femoris) is not palpable. This ligament is taut when the hip is in adduction, and relaxes during hip abduction and during a hip traction mobilization.

Bone movement and axes

- Flexion - extension: around a transverse axis through the head of the femur
- Abduction - adduction: around a sagittal axis through the head of the femur
- Internal - external rotation: around a longitudinal axis through the head of the femur and the knee joint

End feel

- Firm

Joint movement (gliding)

- Convex Rule

Treatment plane

The deep, spherical concave contour of the acetabulum functionally has multiple treatment planes, depending on the position of the joint and the direction of the mobilization force.

- One treatment plane for the hip lies on the concave surface of the weight-bearing portion of the superior acetabulum; i.e., distal traction mobilization is at a right angle to this treatment plane.
- Another treatment plane lies on the concave surface of the anterior-lateral facing portion of the acetabulum; i.e., lateral traction mobilization is at a right angle to this treatment plane.

Zero position

- Thigh in the frontal plane
- The following two lines lie at right angles to each other:
  ... between the anterior superior iliac spine and patella
  ... between the two anterior superior iliac spines

Resting position

- Hip flexed approximately 30°, abducted approximately 30° and slightly laterally rotated

Close-packed position

- Maximal extension and medial rotation and abduction

Capsular pattern

- Medial rotation - extension - abduction - lateral rotation

Hip examination scheme

(Refer to Chapter 3 and 4 for more information on examination)

Tests of function

1. **Active and passive movements, including stability tests and end-feel**

   Transverse axis

Flexion	130 °
Extension from zero	15 °
Extension in abduction	40 °

   Dorsal-ventral axis

Abduction	45°
Adduction	20°

   Longitudinal axis

Lateral rotation	45°
Medial rotation	40°

2. **Translatoric joint play movements, including end-feel**

   Traction - compression (Figure 72a)

   Gliding

   Lateral (Figure 74a)

Hip

3. Resisted movements

	OTHER FUNCTIONS
Flexion	
Iliopsoas	Lateral rotation
Rectus femoris	Knee extension
Tensor fascia latae	Abduction, internal rotation; knee extension, external rotation
Extension	
Gluteus maximus	Lateral rotation
Biceps femoris	Knee flexion, external rotation
Semimembranosus	Knee flexion, medial rotation
Adductor magnus	Adduction
Abduction	
Gluteus medius, minimus	Medial rotation
Tensor fascia latae	Flexion, medial rotation; knee extension, external rotation
Adduction	
Adductor magnus	Medial rotation
Adductor longus, brevis	Flexion, lateral rotation
Pectineus	Flexion, lateral rotation
Gracilis	Flexion, medial rotation
Lateral rotation	
Iliopsoas	Flexion
Gluteus maximus, minimus	Extension
Obturators, gemelli	Extension
Quadratus femoris	Flexion, adduction
Piriformis	Extension, abduction
Medial rotation	
Gluteus minimus, medius	Abduction
Adductor magnus	Adduction
Tensor fascia latae	Flexion, abduction; knee extension, external rotation

4. Passive soft tissue movements

Physiological

Accessory

5. Additional tests

Trial treatment

Traction (Figure 72b)

Hip

■ Hip techniques

Hip distal traction

for pain and hypomobility

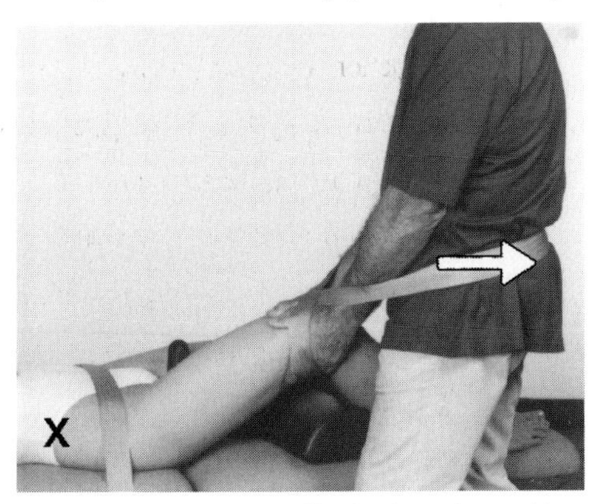

Figure 72a – test and mobilization in resting position

■ Figure 72a: test and mobilization in resting position

Objective

- To evaluate the quantity and quality of distal traction joint play in the hip, including end-feel. Distal traction movement tests the weight bearing area on the upper surface of the acetabulum.
- To decrease pain or increase range-of-motion in the hip.

Starting position

- The patient lies supine.
- Position the hip in its resting position.

Hand placement and fixation

- **Fixation:** To test movement without end-feel, no fixation is required.
- To test end-feel, and for mobilization treatment, use a pommel or stirrup around the right ischial tuberosity to prevent caudal movement of the right innominate; use a strap around the pelvis just below the anterior superior iliac spines to prevent side-bending of the spine.
- **Therapist's moving hands:** Grip around the distal thigh with both hands; use a traction strap over your hands and around your body to reinforce your grip for longer treatments; adjust the strap so that when the strap is taut your arms are nearly straight.

Procedure

- Shift your body backward and pull through your extended arms to apply a Grade I, II, or III distal traction movement.

Hip distal traction
for hypomobility (alternate technique)

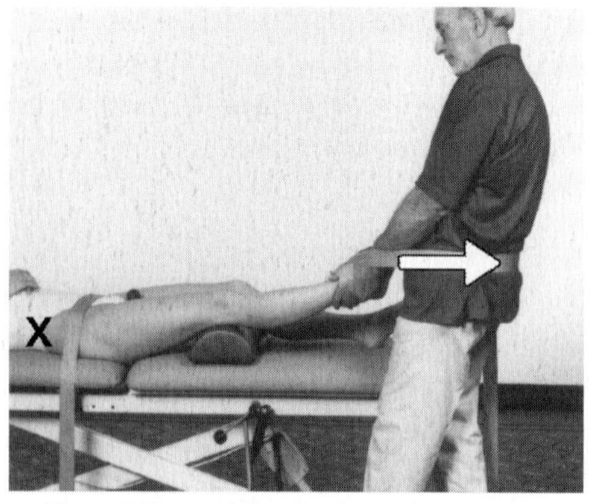

Figure 72b – mobilization in resting position

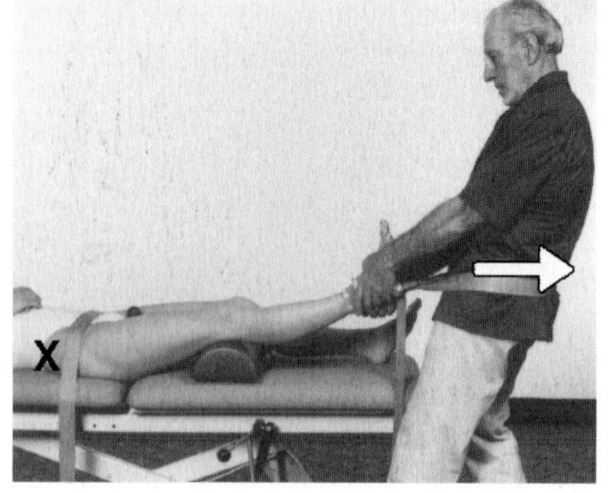

Figure 72c – mobilization in resting position

■ Figure 72b: mobilization in resting position, alternate technique

Objective

- To increase range-of-motion in the hip. Distal traction movement primarily affects the weight bearing area on the superior surface of the acetabulum.

Starting position

- The patient lies supine.
- Position the hip in its resting position.

Hand placement and fixation

- **Fixation:** Use a pommel or stirrup around the right ischial tuberosity to prevent caudal movement of the right innominate. Use a strap around the pelvis just below the anterior superior iliac spines to prevent side-bending of the spine.
- **Therapist's moving hands:** Grip above the patient's ankle with both hands; use a traction strap over your hands and around your body to reinforce your grip for longer treatments; adjust the strap so that when the strap is taut your arms are straight

Procedure

- Shift your body backward and pull through your extended arms to apply a Grade III distal traction movement; pull slowly and sustain each pull for a minute or more.

Contraindication

- Select an alternate technique in the presence of knee pain in knee extension (close-packed position for the knee).

■ Figure 72c: mobilization in resting position, alternate technique

- Use a traction cuff to facilitate your grip for Grade III distal traction mobilization treatments of longer duration.

Hip distal traction

for restricted flexion

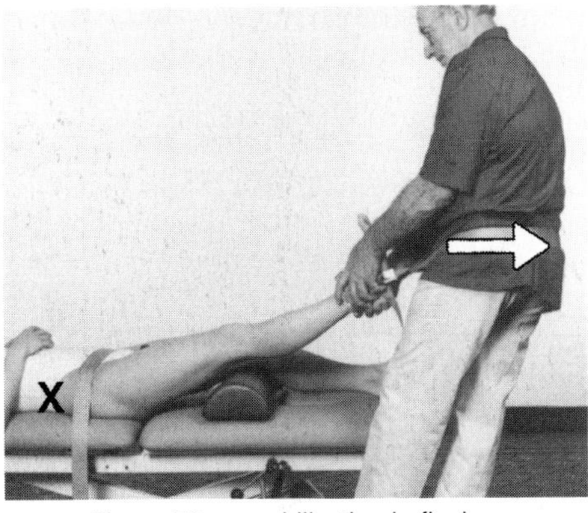

Figure 73a – mobilization in flexion

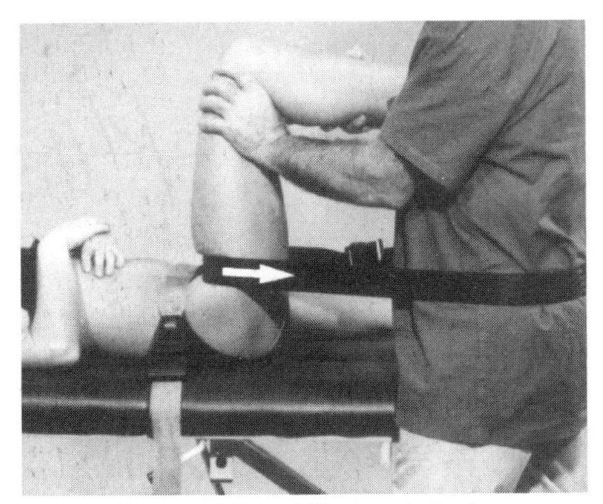

Figure 73b – mobilization in flexion

■ Figure 73a: Flexion progression

Objective

- To increase flexion range-of-motion in the hip by increasing hip distal traction joint play.

Starting position

- The patient lies supine.
- Position the hip near its end range-of-motion into flexion; adjust the height of the treatment surface to control the amount of hip flexion.

Hand placement and fixation

- **Fixation:** Use a pommel or stirrup around the right ischial tuberosity to prevent caudal movement of the right innominate; use a strap around the pelvis just below the anterior superior iliac spines to prevent side-bending of the spine.
- **Therapist's moving hands:** Grip above the patient's ankle with both hands; use a traction strap over your hands and around your body to reinforce your grip for longer treatments; adjust the strap so that when the strap is taut your arms are straight.

Procedure

- Shift your body backward and pull through your extended arms to apply a Grade III distal traction movement; pull slowly and sustain each pull for a minute or more.

■ Figure 73b: Flexion progression for 90° or more

- Position the hip near its end range-of-motion into hip flexion; position the mobilization strap around the patient's proximal thigh and around your body; apply a Grade III distal traction movement to the femur.

Hip distal traction
for restricted extension

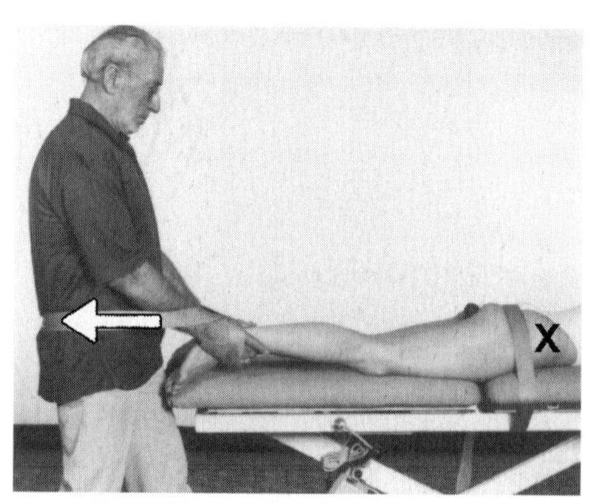

Figure 73c – mobilization in extension

■ Figure 73c: Extension progression

Objective

- To increase extension range-of-motion in the hip by increasing hip distal traction joint play.

Starting position

- The patient lies prone.
- Position the hip near its end range-of-motion into extension.

Hand placement and fixation

- **Fixation:** Use a pommel or stirrup around the right ischial tuberosity to prevent caudal movement of the right innominate; use a strap around the buttocks to prevent side-bending of the spine.
- **Therapist's moving hands:** Grip above the patient's ankle with both hands; in the presence of knee pain in knee extension, bend the patient's knee slightly and grip above the knee.
- Use a traction strap over your hands and around your body to reinforce your grip for longer treatments; adjust the strap so that when the strap is taut your arms are straight.

Procedure

- Shift your body backward and pull through your extended arms to apply a Grade III distal traction movement; pull slowly and sustain each pull for a minute or more.

Hip lateral traction
for pain and hypomobility

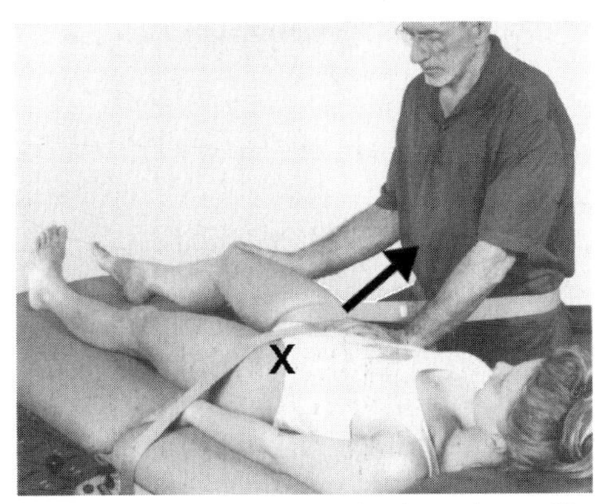

Figure 74a – test and mobilization in resting position

■ Figure 74a: Test and mobilization in resting position

Objective

- To evaluate the quantity and quality of hip lateral traction joint play.
- To decrease pain or to increase range-of-motion in the hip.

Starting position

- The patient lies supine.
- Position the hip in its resting position.

Hand placement and fixation

- **Fixation and therapist's stable hand (left):** Use a strap around and under the patient's pelvis and attach it to the opposite side of the treatment surface; reinforce the strap fixation with your left hand by pressing on the patient's lateral pelvis in a medial and slightly cranial direction.
- **Therapist's moving hand (right):** Use a traction strap around the patient's proximal thigh and around your body; with your right hand, support the patient's knee to control hip resting position.

Procedure

- Shift your body backward and pull through the strap in a lateral and slightly caudal direction to apply a Grade I, II, or III lateral traction movement (in line with the neck of the femur); your right arm and body should move together as one.

Hip lateral traction
for restricted flexion and extension

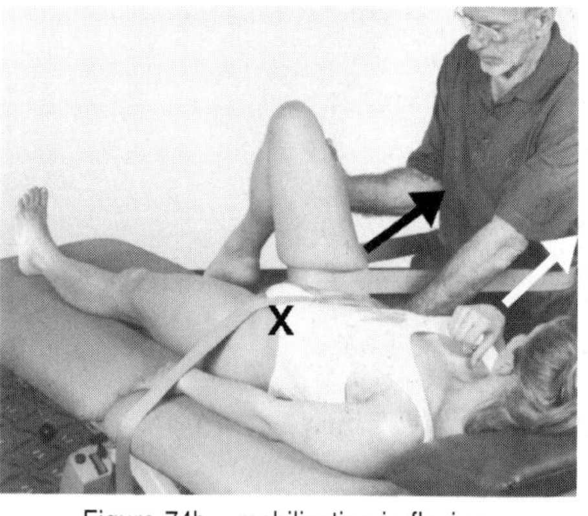

Figure 74b – mobilization in flexion

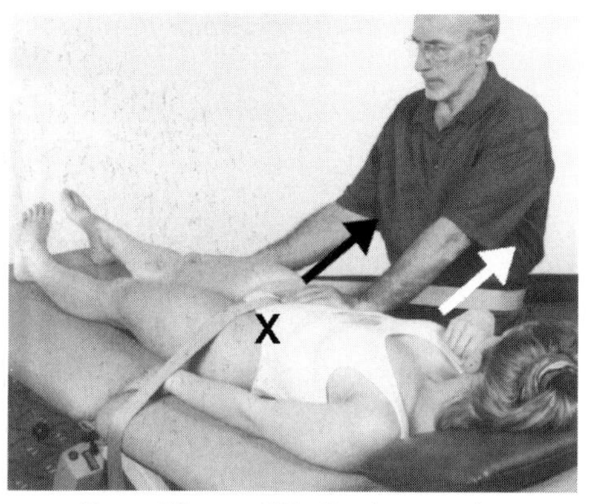

Figure 74c – mobilization in extension

■ Figure 74b: Flexion progression

Objective

- To increase hip flexion and adduction range-of-motion (Convex Rule).

Starting position

- The patient lies supine.
- Position the hip near its end range-of-motion into flexion.

Hand placement and fixation

- **Fixation and therapist's stable hand (left):** Use a strap around and under the patient's pelvis and attach it to the opposite side of the treatment surface; reinforce the strap fixation with your left hand by pressing on the patient's lateral pelvis in a medial and slightly cranial direction.
- **Therapist's moving hand (right):** Use a traction strap around the patient's proximal thigh and around your body; with your right hand, support the patients knee to control hip position.

Procedure

- Shift your body backward and pull through the strap in a lateral and slightly caudal direction to apply Grade III lateral traction movement (in line with the neck of the femur); your right arm and body should move together as one.

■ Figure 74c: Extension progression

- Position the hip near its end range-of-motion in extension and adduction.
- Apply a Grade III lateral traction movement to the femur.
- Note that with greater degrees of extension restriction, the hip may still be in a position of flexion. To increase extension from the zero position, the patient lies prone.

Hip dorsal glide
for restricted flexion

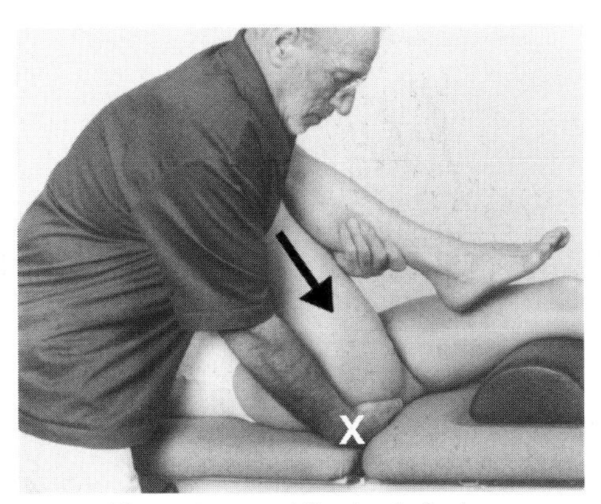

Figure 75a – mobilization in flexion

■ Figure 75a: Flexion progression

Objective

- To increase hip flexion range-of-motion (Convex Rule).

Starting position

- The patient lies supine.
- Position the hip near its end range-of-motion into flexion.

Hand placement and fixation

- **Fixation and therapist's stable hand (right):** The patient's pelvis is fixated against the treatment surface; reinforce the fixation with your right hand beneath the right side of the patient's pelvis.
- **Therapist's moving hand (left):** Hold the patient's leg against your body with your left hand; position the patient's knee against your left shoulder and chest.

Procedure

- Lean your body into the patient's knee along the line of the thigh to apply a Grade III dorsal glide movement; your left arm and body should move together as one.

Hip ventral glide
for restricted extension

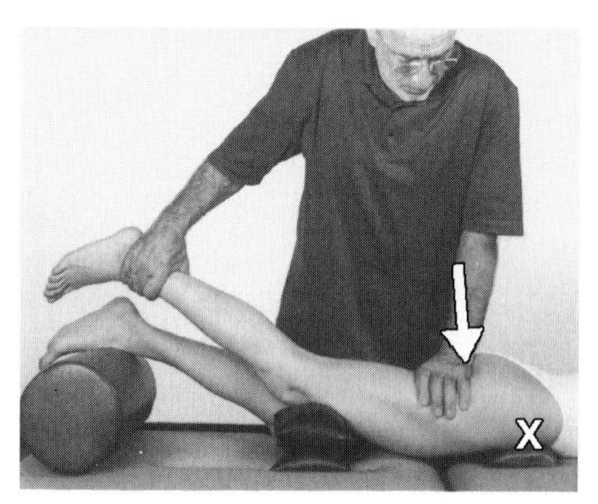

Figure 75b – mobilization in extension

■ Figure 75b: Extension progression

Objective

- To increase hip extension range-of-motion (Convex Rule).

Starting position

- The patient lies supine.
- Position the hip near its end range-of-motion into extension with a firm support above the knee.

Hand placement and fixation

- **Fixation:** The patient's pelvis is fixated against the treatment surface; reinforce the fixation with a sandbag or wedge under the anterior pelvis.
- **Therapist's moving hands:** With your right hand, grip above the patient's ankle to control hip position; with your left hand, grip the posterior thigh just distal to the hip joint space.

Procedure

- Lean over your extended left arm to apply a Grade III ventral glide movement. Your body and both your hands should move together as one.

APPENDIX

■ Upper extremity joint and muscle chart

MUSCLE	NERVE	ROOT	Shoulder *					
			90 ABD	40 ADD	65 FLEX	35 EXTN O	90 INT ROT	60 EXT ROT
Pectoralis major	Pectoral (lat., med.)	C5 – T1		X	X		X	
Pectoralis minor	Pectoral (lat., med.)	C6, 7, 8						
Serratus anterior	Long thoracic	C5, 6, 7						
Trapezius	Accessory C1-C2	C2, 4						
Latissimus dorsi	Thoracodorsal	C6, 7, 8		X		X	X	
Rhomboid	Dorsal scapular	C4, 5						
Levator scapulae	Dorsal scapular	C3, 4, 5						
Deltoid	Axillary	C5, 6	X	1	X	X	1	1
Supraspinatus	Suprascapular	C4, 5, 6	X					1
Infraspinatus	Suprascapular	C5, 6						X
Teres minor	Axillary	C5, 6		1		1		X
Subscapularis	Subscapular	C5, 6, 7		1		1	X	
Teres major	Lower subscapular	C6, 7		X		X	X	
Coracobrachialis	Musculocutaneus	C5, 6, 7		1	X			
Biceps brachii	Musculocutaneus	C5, 6	1		1		1	
Brachialis	Musculocutaneus	C5, 6						
Triceps brachii	Radial	C7, 8		1		1		
Anconeus	Radial	C7, 8, T1						
Pronator teres	Median	C6, 7						
Pronator quadratus	Median	C8, T1						
Flexor carpi radialis	Median	C6, 7						
Palmaris longus	Median	C7, 8						
Flexor carpi ulnaris	Ulnar	C7, 8						
Flexor digit. superf.	Median	C7, 8, T1						
Flexor digitorum profundus	Median, ulnar	C8, T1						
Flexor pollicis longus	Median	C8, T1						
Brachioradialis	Radial	C5, 6, 7						
Extensor carpi radialis (2)	Radial	C6, 7, 8						
Extensor digitorum	Radial	C7, 8						
Extensor digiti minimi	Radial	C7, 8						
Extensor carpi ulnaris	Radial	C7, 8						
Supinator	Radial	C5, 6						
Abductor pollicis longus	Radial	C7, 8						
Extensor pollicis (2)	Radial	C7, 8						
Extensor indicis	Radial	C7, 8						
Lumbricals (III-IV)	Ulnar	C8, T1						
Lumbricals (I-II)	Median	C8, T1						
Adductor pollicis	Ulnar	C8, T1						
Abductor pollicis brevis	Median	C8, T1						
Opponens pollicis	Median	C8, T1						
Flexor pollicis brevis, lat.	Median	C8, T1						
Flexor pollicis brevis, med.	Ulnar	C8, T1						
Interossei	Ulnar	C8, T1						
Hypothenar muscles	Ulnar	C8, T1						

X = prime mover or main function   1 = accessory mover or secondary function   * = with fixated scapula

Shoulder Girdle				Elbow		Forearm		Wrist				Finger				
45	7	30	20	150		80	90	80	90	30	20					
ELEV	DEPR	PROTR	RETR	FLEX	EXT	PRON	SUP	DORS	PALM	ULN	RAD	FLEX	EXT	ABD	ADD	OPP
		X														
	1	1														
	X	X														
X	X		X													
	X		1													
1			X													
X																
				X			X									
				X												
					X											
					1											
				1		X										
						X										
				1		1			X		X					
				1					1							
				1					X	X						
				1					X							
									X			X				
									X			X				
				X		1	1									
				1			1	X			X					
				1				X					X			
				1				1					X			
				1				X		X						
					1		X									
							1		1		1			X		
								1			1		X			
								1			1		X			
												X	1			
												X	1			
															X	1
														1		1
															1	X
												X				1
												X			1	1
												1		1	X	X
												1		X		X

X = prime mover or main function   1 = accessory mover or secondary function

■ Lower extremity joint and muscle chart

MUSCLE	NERVE	ROOT	Hip					
			130 FLEX	15 EXT O	45 ABD	20 ADD	45 EXT ROT	40 INT ROT
Iliopsoas	Lumbar plexus, femoral	L1, 2, 3	X				X	
Gluteus maximus	Inferior gluteal	L5 – S2		X	1	1	X	
Gluteus medius	Superior gluteal	L5, S1	1	1	X		X	1
Gluteus minimus	Superior gluteal	L5, S1	1	1	X			X
Tensor fascia latae	Superior gluteal	L4, 5	1		1			1
Piriformis	Sacral plexus	S1, 2		1	1		X	
Obturator externus	Obturator	L3, 4				1	X	
Obturator internus, gemelli	Sacral plexus	L5, S1		1			X	
Quadratus femoris	Sacral plexus	L5, S1	1			1	X	
Pectineus	Obturator, femoral	L2, 3	1			X	1	
Adductor longus, brevis	Obturator	L2, 3, 4	1			X	1	
Adductor magnus	Obturator, sciatic	L2, 3, 4	1	1		X	1	X
Sartorius	Femoral	L2, 3	1		1		1	
Rectus femoris	Femoral	L2, 3, 4	X		1		1	
Vasti	Femoral	L2, 3, 4						
Gracilis	Obturator	L2, 3	1			X		1
Biceps femoris	Sciatic	L5 – S2		X		1		
Semitendinosus	Tibial	L5 – S2		1		1		
Semimembranosus	Tibial	L5 – S2		X		1		
Gastrocnemius	Tibial	S1, 2						
Popliteus	Tibial	L4 – S1						
Plantaris	Tibial	S1, 2						
Soleus	Tibial	S1, 2						
Tibialis anterior	Deep peroneal	L4, 5						
Extensor hallucis longus	Deep peroneal	L5, S1						
Extensor digitorum longus	Deep peroneal	L5, S1						
Peronei	Superficial peroneal	L5 – S2						
Flexor hallucis longus	Tibial	S2, 3						
Flexor digitorum longus	Tibial	S2, 3						
Tibialis posterior	Tibial	L4, L5						

X = prime mover or main function   1 = accessory mover or secondary function

The charts

We made some compromises to combine both joint and muscle functions in one chart. Where necessary we used the signs "X" and "1" and gave preference to joint function.

Joint Chart: Read from top to bottom. Muscles acting on the joint are listed for each anatomical movement including prime and accessory movers.

Muscle Chart: Read from left to right. Functions are listed for each muscle's main and secondary functions.

Knee				Ankle				Toe	
160	5	45	15	20	40	40	20		
FLEX	EXTN O	EXT ROT	INT ROT	DORS	PLANT	INV	EV	FLEX	EXTN
	1	1							
1			1						
	X								
	X								
1			1						
X		X							
X			1						
X			X						
1					X	1			
1			1						
1					1	1			
					X	1			
				X					
1		1			X				
				1			1		X
					1			X	
					1	1		X	
					1	1		X	
					1	X			

X = prime mover or main function 1 = accessory mover or secondary function

■ Convex-concave table

This table will help you use the Kaltenborn Convex-Concave Rule. Since one usually moves the distal joint partner when testing and mobilizing joints, we list the shape of the distal moving bone in the table.

Joint	Function	Moving bone	Shape
Fingers (PIP, DIP)	flexion / extension	distal phalanx	concave
MCP	abduction / adduction	proximal phalanx	concave
First CM (thumb)	flexion / extension	metacarpal	concave
	abduction adduction	metacarpal	convex
Wrist	dorsal / palmar	capitate, scaphoid, lunate, triquetrum	convex
	dorsal / palmar	trapezii	concave
Radio-ulnar			
Distal	pronation / supination	radius	concave
Proximal	pronation / supination	radius	convex
Humeroradial	flexion / extension	radius	concave
Humero-ulnar	flexion / extension	ulna	concave
Shoulder	all movements	humerus	convex
Sternoclavicular	elevation / depression	clavicle	convex
	protraction / retraction	clavicle	concave
Acromioclavicular	all movements	scapula	concave
Toes (PIP, DIP)	flexion / extension	distal phalanx	concave
MTP	abduction / adduction	proximal phalanx	concave
Foot	all movements	navicular, cuneiform	concave
	inversion / eversion	cuboid	convex
Talocalcaneal	inversion / eversion	anterior calcaneus	concave
		posterior calcaneus	convex
Ankle	dorsal / plantar flexion	talus	convex
Tibiofibular	all movements	fibula head	concave
Knee	all movements	tibia	concave
Hip	all movements	femur	convex
Jaw (TMJ)	all movements	mandible	convex

References

Brodin, H.: *Fysioterapi I.* Studentlitteratur AB, Lund 1979.

Brodin, H., and Moritz, U.: *Fysioterapi II.* Studentlitteratur AB, Lund 1978.

Bromann-Hjortsjö: *Människans rörelsesapparat*, Glerup, Lund 1967. (Figs. 37 and 38 plus the text originally from this book.)

Chapchal, G.: *Grundriss der orthopädischen Krankenuntersuchung*, Enke, Stuttgart 1971.

Cyriax, J.: *Textbook of Orthopaedic Medicine*, Cassell, London 1982.

Derbolowsky, U.: *Leitfaden für Chirotherapie und Manuelle Medizin*, Verlag für Medizin Dr. Ewald Fischer, Heidelberg 1979.

Evjenth, O. and Hamberg, J.: *Muscle Stretching in Manual Therapy*, Alfta Rehab Förlag, 82200 Alfta, Sweden.

Kaltenborn, F. and A. Sivertsen, E. Hansen, J. Bastiansen, O. Aass, R. Gustavsen, H. Fröseth, O. Hagen, O. Bakland, R. Stensnes: *Artikkelsamling*. Frigjöring av exstremitetsledd, Fysioterapeuten, Oslo 1960.

MacConaill, M.A. and Basmajian, J.F.: *Muscles and Movements*, Krieger, Huntington, New York 1977.

Mennell, J.: *The Science and Art of Joint Manipulation. Vol. I.*, Churchill Ltd., London 1949.

Stoddard, A.: *Manual of Osteopathic Technique*, Hutchinson, London 1980.

Anatomical references

Andreassen, E.: *Bevaegeapparatets anatomi*, Gyldendal, Köbenhavn 1976.

Dahl, Olsen, Rinvik: *Menneskets anatomi*, Cappelen, Oslo 1976.

Gray's Anatomy 35th edition, Norwich, Great Britain 1978.

Spalteholz-Spanner: *Handatlas der Anatomie des Menschen.* Bohn, Scheltema & Holkema, NL-Utrecht 1976.

Personal information from

Earlier principals: I. Alvik, V.F. Koefoed, and H. Seyffarth, Oslo.

Former teachers: James Cyriax and James Mennell, London.

Teachers in the Osteopathic College and School in London, especially Alan Stoddard.

· Distraction –
can consider 2 diff. ones
Grade II – measures entire capsule
Grade III – periarticular structures & capsule

telling you it might be limited – how much range/long it takes to get to end pt. compared to o side
– Go to glide to to find out where it's limited w/in jt. – if limited in capsular

· Examine jt. capsule
Grades I – V
" III to examine capsule
· for end feel
; hitting capsul & periarticular struct
;